A Conceptual Guide to Thermodynamics

A Conceptual Guide to Thermodynamics

BILL POIRIER
Texas Tech University

WILEY

This edition first published 2014

Registered office
John Wiley & Sons Ltd, The Atrium, Southern Gate, Chichester, West Sussex, PO19 8SQ, United Kingdom

For details of our global editorial offices, for customer services and for information about how to apply for permission to reuse the copyright material in this book please see our website at www.wiley.com.

Library of Congress Cataloging-in-Publication Data

Poirier, Bill, author.
A conceptual guide to thermodynamics / Bill Poirier.
pages cm
Includes index.
ISBN 978-1-118-84053-5 (pbk.)
1. Thermodynamics. I. Title.
QC311.P757 2014
536′.7–dc23

2014024539

A catalogue record for this book is available from the British Library.

ISBN: 9781118840535

Set in 10/12pt TimesNewRoman by Aptara Inc., New Delhi, India.

Printed and bound in Malaysia by Vivar Printing Sdn Bdh.

1 2014

To students everywhere

Contents

Preface

Thank you for your interest in *A Conceptual Guide to Thermodynamics*. This book is itself a new concept of sorts, which merits some explanation.

First, a description of what this book is *not*. It is *not* a textbook; the discussion is insufficiently complete to serve as the primary text for an undergraduate thermodynamics course, and there are no problems or exercises. Neither is it a popular science or lay person's introduction; the primary intended audience is science and engineering students. Nor is it a history of thermodynamics; though that is itself a fascinating subject, you will find little such discussion here. It is definitely not a book written to impress academic colleagues; they will not be impressed.

What this book *is* is a conceptual and practical guide—a companion to your primary thermodynamics textbook, meant to supplement and clarify the latter. The goal is to simultaneously improve both your fundamental understanding of the material (the "conceptual" part) and your homework and exam performance (the "practical" part), to better "get you through" your thermodynamics course. Culling from over a decade of experience teaching undergraduate physical chemistry thermodynamics at Texas Tech University, this book was written from top to bottom with the practical needs of *you*, the student, foremost in mind.

But why should you buy this (fairly inexpensive) supplement *in addition* to the (no doubt much more expensive) required textbook you have likely already purchased? There are several reasons. First, some textbooks (and some lecturers) may give short shrift to the explication of core thermodynamics concepts such as equilibrium and entropy. The likely reason is clear: there is much material to cover, and they do not want to get bogged down in lengthy explanations and potentially confusing subtleties. Some of the problems arising in this field are indeed profound and intractable; several of its brilliant but frustrated early founders ended their own lives (see Appendix A)… That said, I have learned over my years of teaching thermodynamics that dedicating a modest amount of time during the early stages to a careful (but not *too* rigorous) discussion of the key concepts—if done succinctly and clearly—can lead to major practical benefits for students later on.

Second, a principal advantage of this approach is that the core concepts are pretty much the same across all of the many disciplines that (with good reason!) require thermodynamics training as part of their degree plans. Thus, students of chemistry, physics, biology, geosciences, and the engineering fields, may all benefit from this book, even though the application of this fundamental science varies greatly from field to field. To this end, discipline-specific material is mostly avoided here, in favor of instruction designed to convey the general *logic* of how to solve thermodynamics problems. In this context, memorization *per se* does not really help so much, though many students are naturally inclined to fall back on this tried-and-true companion. In contrast, a conceptual understanding offers something that most students ultimately find to be far more valuable—a sense of how to approach any given problem, as opposed to that uncomfortable state of having "no clue where to begin."

Third, in its role as a true "supplement" to your primary textbook, this book provides explicit references to the latest editions of all of the major thermodynamics texts used by each of the various disciplines listed above. A comprehensive list is provided on p. xv of this book, in the Textbook Guide section. In that section, also, terminology and notation differences between your primary text and this supplement are "translated" for your convenience. Moreover, at the start of each chapter, you will find a map that directs you to the page numbers in your primary text where corresponding material is presented. You will also find the occasional textbook-specific commentary sprinkled throughout this book. All of this is to make it as beneficial and easy for you to use as possible.

Among the range of individuals who would find this book useful, then, one might encounter:

- a **premed student preparing for the MCAT**, for whom thermodynamics is the "hardest class they ever took," but who nevertheless needs a good MCAT score to get into medical school.
- a **brilliant physics major**, who has no trouble solving problems, but is dissatisfied with vague unscientific descriptions such as "entropy is a measure of disorder."
- a graduate or graduate-bound **engineering student**, keen on understanding what is really going on in real-world applications.
- a **geochemist** or **materials scientist**, seeking a better intuition about the role of free energy in amorphous solids, or about processes that take place far from standard temperatures and pressures.
- anyone else—student, teacher or research professional—who would benefit from a better understanding of this interesting subject.

Acknowledgments

I am extraordinarily grateful for the substantial help and encouragement I have received from various individuals, in the course of putting together this book. I must thank Texas Tech University, and especially Carol Korzeniewski, the Chair of Chemistry and Biochemistry, for supporting my faculty development leave in the fall of 2011. Much of this book was designed during this "sabbatical" period, and might not have been possible—or at least severely delayed—without it. I must also thank Gérard Parlant and the Centre national de la recherche scientifique at the Université Montpellier 2, for acting as my hosts during this period (and particularly Odile Eisenstein, for the use of her office). My wife, Anne, is also to be commended for her copious patience and understanding—particularly considering the other ways we might have spent our time in the *sud de France*. I thank David Tannor and Michael New for the LaTeX style files used to typeset the earliest versions of the manuscript, and also Peter Wilson and Piet van Oostrum—respective authors of the `changepage` and `fancyhdr` packages, used extensively for the later versions. Of course, I am grateful to the many colleagues, educators, formal reviewers, and editors who read through those preliminary drafts, and provided much useful feedback. Above all, however, I must fervently acknowledge the many students who have urged me over the years to write such a book—without their ardor and persistent arm-twisting, it never would have happened.

Saint Mathieu de Tréviers, France

Bill Poirier
December, 2011

As this book nears completion, it has become clear that I must acknowledge by name those tireless volunteers who have reviewed the entire book manuscript; their copious and insightful "notes" have resulted in an immeasurably superior final product. To Joseph Ellis, Thomas Gibson, Jeremy Maddox, Jason L. McAfee, Michele S. McAfee, Corey Petty, and Caroline Taylor, you have earned my eternal gratitude. I would also be remiss not to acknowledge two specific editors at John Wiley & Sons, who bent over backwards and went the extra mile, time and again; this book *certainly* would not have happened without the many efforts taken on its behalf by Sarah Higginbotham and Sarah Keegan.

The quotation on p. 78 is reprinted with permission from Y. A. Çengel and M. A. Boles, *Thermodynamics: An Engineering Approach*, seventh edition. Copyright © 2011 by McGraw-Hill Education. The quotations on pp. 79 and 99 are from RAFF, LIONEL M., PRINCIPLES OF PHYSICAL CHEMISTRY, 1st Edition, © 2001, pp, 170, 144. Reprinted by permission of Pearson Education, Inc., Upper Saddle River, NJ.

This book was fueled not so much by caffeine as by YES; may they make it into the Rock and Roll Hall of Fame.

Vienna, Austria

Bill Poirier
October, 2013

Textbook Guide

Thermodynamics is a cornerstone of many scientific disciplines. As such, there are many different textbooks that address this important subject—and a corresponding myriad of conventions, terminologies, and notations used. This section is designed to sort all of that out—at least for the specific reference textbooks considered here, listed (by discipline) below.

The list includes the latest edition of the most commonly used texts. Please do not be too discouraged if you do not see your primary text listed below; given the multiple disciplines represented, it is an inevitability that many excellent textbooks have to be omitted. That said, a *greatly expanded* list of reference texts is available on the companion website (`http://www.conceptualthermo.com`), which you should also consult. There, you will find customized, textbook-specific materials for the older and (as they are released) newer editions of the textbooks listed below—as well as for many other texts that are less frequently used.

If your favorite book is not listed on the website, please feel free to let me know via the website itself, or by sending an email directly to `feedback@conceptualthermo.com`. Please send detailed information (including ISBN) for the textbook that you are using, so that I can incorporate it into future book materials. In any case, you can still learn from the conceptual explanations provided in this book, without necessarily having to own *any* thermodynamics textbook.

Whether you wind up loving or hating this book, or somewhere in between, I certainly would appreciate your feedback—especially in the form of helpful suggestions for making it better. You can submit these via the website and email address above. If you like the book, consider spreading the word—to professors, social media friends, and even real friends, as appropriate.

0.1 List of Thermodynamics Textbooks by Discipline

In the list below, the boldfaced word at the start of each bibliographic entry is the **keyword** that will be used throughout the rest of this book to refer to that specific reference textbook. (In all but one case, the keyword is simply the first author's last name.) When looking up your primary textbook here, be sure to pay close attention to the *edition number* and publication year.

Thermodynamics Textbooks

Biological Sciences:

Atkins-life: P. Atkins and J. de Paula, *Physical Chemistry for the Life Sciences*, second edition (W. H. Freeman, 2011).

Chang: R. Chang, *Physical Chemistry for the Biosciences* (University Science Books, 2005).

Tinoco: I. Tinoco, Jr., K. Sauer, J. C. Wang, J. D. Puglisi, G. Harbison, and D. Rovnyak, *Physical Chemistry: Principles and Applications in Biological Sciences*, fifth edition (Prentice-Hall, 2013).

Chemistry:

Atkins: P. Atkins and J. de Paula, *Physical Chemistry*, ninth edition (W. H. Freeman, 2010).

Engel: T. Engel and P. Reid, *Physical Chemistry*, third edition (Pearson, 2013).

Levine: I. N. Levine, *Physical Chemistry*, sixth edition (McGraw-Hill, 2009).

McQuarrie: D. A. McQuarrie and J. D. Simon, *Physical Chemistry* (University Science Books, 1997).

Silbey: R. J. Silbey, R. A. Alberty, and M. G. Bawendi, *Physical Chemistry*, fourth edition (John Wiley & Sons, Inc., 2005).

Engineering:

Cengel: Y. A. Çengel and M. A. Boles, *Thermodynamics: An Engineering Approach*, seventh edition (McGraw-Hill, 2012).

Elliot: J. R. Elliott and C. T. Lira, *Introductory Chemical Engineering Thermodynamics*, second edition (Prentice-Hall, 2012).

Moran: M. J. Moran, H. N. Shapiro, D. D. Boettner, and M. B. Bailey, *Fundamentals of Engineering Thermodynamics*, seventh edition (John Wiley & Sons, Inc., 2012).

Prausnitz: J. M. Prausnitz, R. N. Lichtenthaler, and E. Gomes de Azevedo, *Molecular Thermodynamics of Fluid-Phase Equilibria*, third edition (Prentice-Hall, 1999).

Sandler: S. I. Sandler, *Chemical, Biochemical, and Engineering Thermodynamics*, fourth edition (John Wiley & Sons, Inc., 2006).

Smith: J. M. Smith, H. C. Van Ness, and M. M. Abbott, *Introduction to Chemical Engineering Thermodynamics*, seventh edition (McGraw-Hill, 2005).

Geosciences:

Anderson: G. Anderson, *Thermodynamics of Natural Systems*, second edition (Cambridge University Press, 2005).

Faure: G. Faure, *Principles and Applications of Geochemistry*, second edition (Prentice-Hall, 1998).

Physics:

Baierlein: R. Baierlein, *Thermal Physics* (Cambridge University Press, 1999).

Callen: H. B. Callen, *Thermodynamics and an Introduction to Thermostatistics*, second edition (John Wiley & Sons, Inc., 1985).

Kittel: C. Kittel and H. Kroemer, *Thermal Physics*, second edition (W. H. Freeman, 1980).

Reif: F. Reif, *Fundamentals of Statistical and Thermal Physics* (McGraw-Hill, 1965; reprinted by Waveland Pr, 2008).

Schroeder: D. Schroeder, *An Introduction to Thermal Physics* (Addison-Wesley, 2000).

0.2 Terminology and Notation Used in This Book

The terminology and mathematical notation of the most important physical, chemical, and thermodynamic quantities that are employed in this book are listed in the table below. The page on which each quantity is first introduced or defined is also listed.

Terminology and Notation

Quantity	Symbol	Page Number
Amount of Substance (number of moles)	n	18
Amount of Substance (number of particles)	N	18
Avogadro's Number	N_A	18
Boltzmann Constant	k	30
Chemical Potential	μ	125
Compressibility Factor	Z	30
Efficiency (heat engine)	η	110
Energy (molecular)	E	34
Enthalpy	H	68
Entropy	S	84
Exergy	E	121
Expansion Coefficient	α	71
Force	F	28
Fugacity	f	126
Gas Constant	R	30
Generic Thermodynamic Quantity (molar)	X_m	19
Generic Thermodynamic Quantity (surroundings)	X_sur	50
Generic Thermodynamic Quantity (system)	X	19
Generic Thermodynamic Quantity (total system)	X_tot	50
Gibbs Free Energy	G	117
Heat	Q	59
Heat Capacity at Constant Pressure	C_P	68
Heat Capacity at Constant Volume	C_V	67
Heat Capacity Ratio (adiabat coefficient)	γ	70
Helmholtz Free Energy	A	117
Internal Energy	U	36
Internal Pressure	π_T	70
Isothermal Compressibility	κ_T	71
Mass (per mole)	M	18
Mass (per particle)	m	18
Number of Available Molecular States	Ω	81
Position	(x, y, z)	20
Pressure	P	17
Temperature	T	39
Velocity	(v_x, v_y, v_z)	20
Volume	V	17
Work (expansion)	W	58

0.3 Terminology and Notation Used in Textbooks

Consult the key below to "translate" the terminology and notation of this book into that of your primary textbook (or to the **IUPAC** Gold Book standard). The key is easy to use. Textbooks are listed by keyword, in alphabetical order. Unless stated otherwise, the notation used here, and that of a given text, are presumed to be identical. Where the two notations differ, that of this book appears to the left of the arrows presented in the key, and that of the textbook to the right.

For example, after the **Atkins** keyword below, the string "$P \to p$" appears. This means that this book uses 'P' to denote the pressure, whereas *Physical Chemistry* by P. Atkins and J. de Paula uses 'p'. Likewise, since W in this book is the work done *on* the system *by* the surroundings, the "$W \to -W$" that appears after **Cengel** below implies that 'W' in that textbook is the work done *on* the surroundings *by* the system (a typical engineering convention).

Sometimes, more than one expression appears to the right or left of a given arrow. When found to the right of the arrow, separated by "or," it means that the textbook uses more than one notation to denote the same mathematical quantity. Likewise, multiple expressions to the left of an arrow, separated by "and," correspond to distinct quantities that the textbook represents using the same notation. When "(sometimes)" appears, it means that the textbook sometimes uses the phrase indicated, and other times uses the terminology of this book.

Note that the Terminology and Notation Key can be downloaded in tabular form, just for your specific textbook, from the website (`http://www.conceptualthermo.com`). For reference purposes, it is recommended that you download and print a copy of this key—together with various other textbook-specific materials that are also available online, for free.

Terminology and Notation Key

Anderson: molar mass $M \to$ gram formula weight gfw; $P_{\text{sur}} \to P_{\text{ext}}$; $Q \to \boldsymbol{q}$; $dQ \to \delta \boldsymbol{q}$; $W \to \boldsymbol{w}$; $dW \to \delta \boldsymbol{w}$; $X \to \mathbf{Z}$; $X_{\text{m}} \to Z$; $\Omega \to W$; molecular state $\to$ microstate; state function $\to$ property or state variable

Atkins: m and $nM \to m$; $P \to p$; $P_{\text{sur}} \to p_{\text{ex}}$; $Q \to q$; $W \to w$; $\Omega \to W$; compressibility factor $\to$ compression factor; ideal gas $\to$ perfect gas; quantity $\to$ property (sometimes)

Atkins-life: $nM \to m$; $P \to p$; $P_{\text{sur}} \to p_{\text{ex}}$; $Q \to q$; $W \to w$; ideal gas $\to$ perfect gas; quantity $\to$ property (sometimes)

Baierlein: $A \to F$; E and $U \to E$; $dQ \to q$; $U \to E$ or $\langle E \rangle$; $dW \to w$; $X_{\text{sur}} \to X_{\text{environment}}$; $X_{\text{tot}} \to X_{\text{total}}$; $\Omega \to$ multiplicity; ideal gas law $\to$ empirical gas law (sometimes); quantity $\to$ attribute or property

Callen: $A \to F$; $E \to \varepsilon$ or E; $f \to (\xi/\xi^{\circ})P^{\circ}$; $k \to k_B$; $n \to N$; $N \to \tilde{N}$; $dQ \to đQ$; $W \to W_M$; $dW \to đW_M$; $\eta \to \varepsilon_e$; diathermic $\to$ diathermal; exact differential $\to$ perfect differential; quantity $\to$ parameter; spontaneous $\to$ irreversible

Cengel: $C_P \to mc_p$; $C_V \to mc_V$; exergy E $\to$ total exergy X; k and $\gamma \to k$; $nM \to m$; $n \to N$; $dQ \to \delta Q$; $R \to R_u$; $(v_x, v_y, v_z) \to (V_x, V_y, V_z)$; $V \to \mathcal{V}$; $W \to -W$; $dW \to -\delta W$; $\alpha \to \beta$; $\kappa_T \to \alpha$; number of available molecular states $\Omega \to$ thermodynamic probability p; closed system $\to$ control mass (sometimes); expansion coefficient $\to$ volume expansivity; open system $\to$ control volume (sometimes); Second Law $\to$ increase of entropy principle (sometimes); state function $\to$ property; work (expansion) $\to$ (moving) boundary work

Chang: $k \to k_B$; m and $nM \to m$; $M \to \mathcal{M}$; $P_{\text{sur}} \to P_{\text{ex}}$ or P_{ext}; $Q \to q$; $dQ \to đq$; $W \to w$; $dW \to đw$; $X \to X_{\text{sys}}$ or X; $X_{\text{m}} \to \overline{X}$; $X_{\text{sur}} \to X_{\text{surr}}$; $X_{\text{tot}} \to X_{\text{univ}}$; $\Omega \to W$; Faraday constant $\to F$; quantity $\to$ property

Elliott: E $\to \underline{B}$; m and $nM \to m$; $(v_x, v_y, v_z) \to (u_x, u_y, u_z)$; $X \to \underline{X}$; $X_\mathrm{m} \to X$; $X_\mathrm{sur} \to \underline{X}_{surr}$; $X_\mathrm{tot} \to \underline{X}_{univ}$; $\alpha \to \alpha_P$ or α; $\eta \to \eta_\theta$; $\Omega \to W$ or p_i; diathermic → diathermal; exergy → availability (sometimes); molar quantity → intensive quantity; spontaneous → irreversible; state function → property; work (expansion) → expansion/contraction work

Engel: E and $\eta \to \varepsilon$; $k \to k_B$; $P_\mathrm{sur} \to P_{external}$; $Q \to q$; $dQ \to đq$; $W \to w$; $dW \to đw$; compressibility factor $Z \to$ compression factor z; $\alpha \to \beta$; $\kappa_T \to \kappa$; $\Omega \to W$; diathermic → diathermal; quantity → function; spontaneous → natural

Faure: $Q \to q$; $U \to E$; $W \to -w$

IUPAC: $C_P \to C_p$; $f \to \overline{p}$ (sometimes); mass (per mole) $M \to$ relative molecular mass M_r; $N_\mathrm{A} \to L$ (sometimes); $P \to p$; $Q \to q$ (sometimes); $W \to w$ (sometimes); $\Omega \to W$; compressibility factor → compression factor

Kittel: $A \to F$; $C_P \to k_B C_p$; $C_V \to k_B C_V$; $E \to \varepsilon$; $F \to f$; $k \to k_B$; $m \to M$; $N_\mathrm{A} \to N_0$; $N/V \to n$; $P \to p$; $dQ \to đQ$; entropy $S \to$ conventional entropy $k_B\sigma$ or S; $kT \to$ fundamental temperature τ or kT; $dW \to đW$; $\mu \to N_0\mu$; $\Omega \to g$; electron rest mass → m; heat capacity → conventional heat capacity; internal energy → energy; molecular state → microstate; Second Law → law of increase of entropy (sometimes); spontaneous → irreversible; surroundings → reservoir; thermodynamic state → macrostate

Levine: m and $nM \to m$; $Q \to q$; $W \to w$; $X \to X_\mathrm{syst}$ or X; $X_\mathrm{sur} \to X_\mathrm{surr}$; $X_\mathrm{tot} \to X_\mathrm{univ}$; $\eta \to e$; $\Omega \to W$; diathermic → thermally conducting; movable wall → nonrigid wall; diffusive equilibrium → material equilibrium; work (expansion) → P–V work; expansion coefficient → thermal expansivity

McQuarrie: $nM \to M$; f and $F \to F$; $k \to k_\mathrm{B}$; $Q \to q$; $(v_x, v_y, v_z) \to (u_x, u_y, u_z)$; $W \to w$; $X_\mathrm{m} \to \overline{X}$; $X_\mathrm{sur} \to X_\mathrm{surr}$; $X_\mathrm{tot} \to X_\mathrm{total}$; $\Omega \to W$

Moran: $A \to \mathbf{\Psi}$; $C_P \to mc_p$; $C_V \to mc_v$; $k \to \mathrm{k}$; $nM \to m$; $P \to p$; $dQ \to \delta Q$; $R \to \overline{R}$; $(v_x, v_y, v_z) \to (\mathrm{V}_x, \mathrm{V}_y, \mathrm{V}_z)$; $W \to -W$; $dW \to -\delta W$; $X \to mx$; $X_\mathrm{sur} \to X]_\mathrm{surr}$; $X_\mathrm{tot} \to X]_\mathrm{isol}$; $\alpha \to \beta$; $\gamma \to k$; $\kappa_T \to \kappa$; $\Omega \to w^N$; closed system → control mass (sometimes); expansion coefficient → volume expansivity; heat capacity ratio → specific heat ratio; open system → control volume (sometimes); state function → property

Prausnitz: $C_P \to nc_p$ or C_p; $C_V \to nc_v$ or C_v; $P_\mathrm{sur} \to P_E$; $T_\mathrm{sur} \to T_B$; $X_\mathrm{m} \to x$; $Z \to z$; $\alpha \to \alpha_p$; $\Omega \to W$; expansion coefficient → thermal expansion coefficient

Reif: $A \to F$; $F \to X$; $n \to \nu$; $N/V \to n$; $N_\mathrm{A} \to N_a$; $P \to p$ or $\bar{p}$; $dQ \to đQ$; $U \to E$ or $\bar{E}$; $W \to -W$; $dW \to -đW$; $\kappa_T \to \kappa$; μ and $M \to \mu$; diathermic → thermally conducting; internal energy → mean energy (sometimes); pressure → mean pressure (sometimes); thermodynamic state → macrostate

Sandler: $C_P \to nC_\mathrm{P}$; $C_V \to nC_\mathrm{V}$; $M \to \mathrm{mw}$; $n \to N$; $nM \to M$; $X_\mathrm{m} \to \underline{X}$; $\kappa_T \to \kappa_\mathrm{T}$ or κ_T; pressure → absolute pressure (sometimes); state function → property; heat capacity ratio → specific heat ratio

Schroeder: $A \to F$; $X_\mathrm{tot} \to X_\mathrm{total}$; $\alpha \to \beta$; $\eta \to e$; compressibility factor → compression factor; expansion coefficient → thermal expansion coefficient; heat capacity ratio → adiabat exponent; internal energy → total energy; work (expansion) → compression work

Silbey: $F \to f$; $P_\mathrm{sur} \to P_\mathrm{ext}$; $Q \to q$; $dQ \to đq$; $W \to w$; $dW \to đw$; $X_\mathrm{m} \to \overline{X}$; $\eta \to \epsilon$; $\kappa_T \to \kappa$; diathermic → heat-conducting; quantity → property (sometimes)

Smith: m and $nM \to m$; $(v_x, v_y, v_z) \to (u_x, u_y, u_z)$; $X \to X^t$; $X_\mathrm{m} \to X$ or MX; $X_\mathrm{sur} \to nX_\sigma$ or nMX_σ; $\alpha \to \beta$; $\kappa_T \to \kappa$; exact differential → infinitesimal change; expansion coefficient → volume expansivity; inexact differential → infinitesimal amount; number of chemical species → N; open system → control volume (sometimes); spontaneous → irreversible; state function → property (sometimes)

Tinoco: $C_P \to C_p$; $k \to k_\mathrm{B}$; m and $nM \to m$; $P \to p$; $Q \to q$; $W \to w$; $X_\mathrm{sur} \to X_\mathrm{ex}$; $\kappa_T \to \kappa$; $\Omega \to W$; non-state function → path variable; state function → state variable; work (expansion) → pV work

Chapter 1

About This Book

Put yourself in the mindset of a 19th-century scientist—austere, heavily-bearded (see Appendix A), and intently focused on the "new" science of thermodynamics. One of your main practical motivations is *steam engines*—i.e., cylinders, pistons, boilers and the like—and specifically, how to improve their performance. It was in this engineering context that the field was born.

It was not called "classical thermodynamics" just yet...

Very quickly, however, the importance of thermodynamics in many other areas became widely recognized. It even gained a strong foothold in the popular culture and imagination of the day (see Appendix B). Today, thermodynamics has developed into a cornerstone subject for virtually every discipline of science and engineering.

Simply put, thermodynamics is the science that addresses how matter behaves at the *macroscopic* scale (i.e., at the ordinary scale of everyday human experience), and also how this behavior relates to the *molecular* scale (i.e., of individual molecules, nanoparticles, etc.) As such, it is a subject of profound practical as well as fundamental importance.

From the Texts: Some authors prefer the terms "*macro*" or "*bulk*."

From the Texts: Some authors prefer the terms "*micro*" or "*nano*."

Thermodynamics is used, for example, to characterize the properties of new materials developed for specific applications—from hydrogen cars to wind turbines to fabrics. Machines built to provide mechanical work, to condition air, or to perform computations, all rely on thermodynamic principles to maximize their efficiency. Likewise, these same thermodynamic principles govern a great many processes observed in nature: from the smallest nanodevices to the whole universe itself; from ultrafast chemical and photophysical reactions to "ultraslow" geological transformations.

not to mention the bioenergetics of living organisms, lying somewhere in the middle...

But our job here is not really to explain to you how important thermodynamics is to your chosen discipline, nor to try to convince you that it's "cool." Our job is to lay a solid conceptual foundation for the subject, by clearly explaining and defining the core concepts (see Section 1.3), and clarifying the subtle—even insidious—distinctions that arise. As when building any stable edifice, success with thermodynamics, both in the classroom and in professional practice, requires a solid foundation. This means attaining both a conceptual and practical mastery of the fundamentals of the subject.

From the Texts: Your primary textbook is there to do that.

A Conceptual Guide to Thermodynamics, First Edition. Bill Poirier.

Companion website: http://www.conceptualthermo.com

Such study may not seem fun and exciting, and at first glance may even appear extraneous to your chosen discipline. Don't be fooled. Whatever your field of study, a careful consideration of the fundamentals early on in your thermodynamics (or related) course will generally lead later on to smoother sailing, less head-banging, and ultimately, better grades.

1.1 Who Should Use This Book?

As might be imagined, the various disciplines that train their students in thermodynamics tend to emphasize quite different aspects of this important subject. Despite this variety, the core thermodynamics concepts are the same across all disciplines; it is this common core that serves as the focus of this book.

As a general rule, if you are taking a thermodynamics or related course, and your primary textbook is listed either on p. xv of the Textbook Guide section, or on the companion website, then you are a good candidate for this supplement. (See the Preface for some specific examples of good candidates.)

On the Website: http://www.conceptualthermo.com

If you are an undergraduate chemistry or physics major, then this book is very well suited for your "Physical Chemistry" or "Thermal Physics" course, respectively. Graduate students in these disciplines who are preparing for their cumulative or qualifying examinations may also find the book to be of benefit, as will premed students preparing for the MCAT.

Chemical engineering majors typically take two semesters of "Chemical Engineering Thermodynamics." This book may help such students, with regard to their understanding of concepts such as free energy and fugacity that play a vital role in phase and reaction equilibria, and in chemical yields. Other engineering majors typically take a one-semester "Engineering Thermodynamics" course. The more highly motivated graduate and graduate-school-bound engineering students may also find the molecular description of entropy useful, particularly as a direct connection with macroscopic entropy is provided here.

Undergraduate geoscience students might use this book in conjunction with their "Geochemistry" course. However, it is probably better suited for graduate students taking a "Chemical Thermodynamics for Geoscientists" course. Likewise, biological science students might find it useful it as part of a "Physical Chemistry for Biological Sciences" course, particularly vis-à-vis the discussion of free energy.

Whether assigned as required reading for a specific course, or purchased voluntarily as supplemental reading, this book is primarily intended for students. That said, instructors may also find the material useful for lecture preparation and course development. Being a supplement, use of this book need not necessitate revision of an existing lesson plan—unlike, say, a change of primary textbook.

Finally, we note that various ebook formats are also available; consult the website for details.

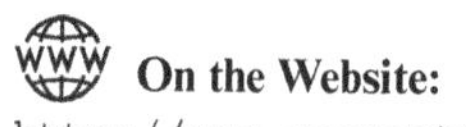

On the Website: http://www.conceptualthermo.com

1.2 Philosophy of This Book

This book is a *supplement*, and decidedly *not* a comprehensive, standalone thermodynamics textbook. As such, the focus is on developing (fairly) precise definitions of the key concepts used in the field, as well as on pointing out common misconceptions, so as to make the subject easier to understand. We thus leave the splashy color graphics—together with a description of the latest "sexy" applications—to your primary textbook, and instead concentrate here on a clear presentation of the core ideas. The premise is that a solid conceptual understanding of the core principles will lead to a better understanding of the subject—and ultimately to greater practical success, e.g., in solving problems. For problems, the main goal is to convey a sense of how these should be initially conceptualized and approached—rather than to work through many detailed examples from start to finish, as you can no doubt already find in your primary textbook. Again, the aim is to complement, rather than to replace, the latter.

From the Texts

From the Texts: A good source of worked problems is *Schaum's Outline of Thermodynamics*, whether *for Engineers* or *With Chemical Applications* (McGraw-Hill).

In the interest of clarity, this book does not include many formal mathematical proofs as such, but tends to rely instead on intuitively plausible arguments. The level of rigor is presumably appropriate for most disciplines at the introductory level. At the least, it should serve to get students past the dangerous practice of just "plugging in the formula" *without really understanding when or why it applies*—which is a major goal of this book.

if not *the* major goal...

In any case, a more rigorous treatment would require *statistical mechanics*—an advanced field lying for the most part beyond the scope of this book. It is true that we do rely on a molecular, or statistical, viewpoint, in the development of some of the important core concepts—notably energy and entropy (see Section 1.3). We do so, however, only to the extent necessary to clarify a conceptual understanding of these vitally important thermodynamic quantities. Most other quantities are derived directly at the macroscopic scale.

though this book could serve as a useful primer for a statistical mechanics-based course...

For additional discussion on the difference between macroscopic and molecular scales, see Chaps. 2 and 3.

1.3 Four Core Concepts of Thermodynamics

Science is—well—an "exact science," and so meaningful scientific discourse depends vitally on knowing precisely what one is talking about. Even so, science often commandeers words and phrases from imprecise, everyday usage—words such as "energy," for instance.

This practice presents both a benefit and a hindrance. On the one hand, it provides us with intuition about the meaning of new scientific terms. On the other hand, it can limit our thinking, or bias us towards certain expectations that are ultimately unwarranted. This situation pervades science in general, but it is *especially* true of thermodynamics. Here, the main concepts can be subtle and confusing—and even inconsistent or ill-defined, historically speaking.

We therefore rely on *definitions*, to clarify the precise meaning of terminology. Here is the first definition that we will encounter in this book, the definition of a definition:

Definition 1.1 *A* definition *is a complete specification of what something is.*

A central goal of this book is to provide unambiguous definitions—together with an intuitive understanding—of four core concepts in the field of thermodynamics. These are: *equilibrium*, *energy*, *entropy*, and *free energy*. Note that there are also four Laws of Thermodynamics, which more or less correlate to the four core concepts. This book is also divided into parts, one of which is also dedicated to each of the four core concepts. Each of these concepts will be properly defined in due course; in the meantime, a very brief introduction is presented here.

equilibrium The word "equilibrium" has certain connotations that are fairly appropriate with regard to its thermodynamic usage. One thinks of stillness and stability. Do *not*, however, imagine that the molecules in an equilibrium system are stationary [***Don't*** **Try It !!**]. Indeed, molecules in "still" air, for example, move approximately as fast as those in an intense tornado.

energy This is the most important quantity in all of science. In thermodynamics, though, the term has at least three distinct meanings (see Chapter 5). One must therefore be careful about which kind of energy—and energy conservation—one is talking about.

entropy This term does not get used so much in everyday speech—or rather, did not, before thermodynamics entered the popular culture (see Appendix B). In this context, "entropy" often suggests "chaos" or "disorder." Within thermodynamics, it remains the most confusing and least understood concept, by far. Suffice it to say that a major goal of this book is to explain clearly and precisely what entropy actually is.

free energy Often described as the "energy available for useful work," free energy (and its engineering cousin, exergy) is the determining factor underlying much of what occurs naturally in the world. Conceptually, it can be described as follows:

$$\text{free energy} = \text{energy} - \text{entropy} \tag{1.1}$$

Thus, free energy implies a *tension* or competition between energy and entropy—these being the two primary forces that drive natural processes forward. Note that free energy can be—and often is—*negative*.

1.4 How to Use This Book

As this book is a supplement, it should be used differently from a textbook. To begin with, it is not expected that you should read the *entire* book, nor should you necessarily read it in the order presented. If it is required reading for a thermodynamics course, your instructor will sort out the details of what you should read, and when. If you have purchased this book on your own, then you should read whichever sections seem appropriate, or otherwise strike your fancy. Note that there is a lot of cross-referencing, so that if at any point you need something from earlier in the book, you will generally be directed there.

To help you identify which sections of your primary text correspond to specific chapters of this book, please refer to the maps given at the start of each chapter, and on the website. Likewise, the Terminology and Notation Key on p. xviii of this book will sort out any differences in that regard, that might otherwise lead to confusion. Also, don't forget to register on the website, which will open up many more resources to you for free, including textbook-specific materials.

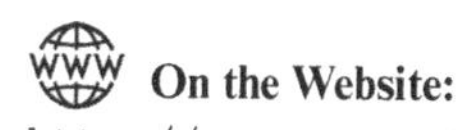

On the Website: http://www.conceptualthermo.com

With the exception of this Introduction and the Appendices, in which we deliberately indulge ourselves a bit, the chapters tend to be short and to the point. Generally, the most important or fundamental material is presented towards the front of each chapter, whereas the more advanced or esoteric material is found in the later sections. Sometimes, *marginal notes* will be used to warn you that such material is being approached. In general, marginal notes refer to the period, comma, or italicized word(s) in the line of the main body of text corresponding to the first line of the marginal note.

such as this one...

Thus, this marginal note refers to "note.", whereas the marginal note above refers to "*marginal notes*".

Your conceptual muscles will also be stretched, in special **To Ponder...** areas. Here is the first To Ponder, referring to Definition 1.1, as presented in Section 1.3:

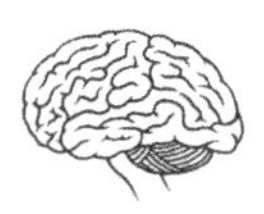

▷▷▷ **To Ponder...** Definition 1.1 allows for the possibility of multiple definitions for the same thing—provided that all such definitions are equivalent. Does this mean that the definition of "definition" could itself have multiple forms?

Occasionally, you will also encounter the dreaded, *double* brains:

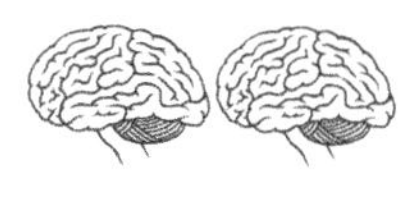

▷▷▷ **To Ponder...*at a deeper level.*** These areas are reserved for the most advanced material, often relying on such subjects as statistical mechanics or quantum mechanics that may lie well beyond the level of a traditional introductory thermodynamics course. Enter at your own risk! Or, work with a study buddy—two brains are better than one!

There are also special **Try It!!** areas:

▷▷▷ **Try It!!** Flip through this book and you will find a number of these, encouraging you to conduct various "experiments" or exercises, and exhorting you to look certain things up in your primary textbook.

Also, because students of thermodynamics tend to make the same kinds of mistakes, special warnings are presented in the ***Don't* Try It!!** areas:

▷▷▷ ***Don't* Try It!!** Here, we warn you about the most common pitfalls that you should try your best to avoid. Heed our advice, and don't become another statistic!

Likewise, we also offer you useful, practical advice, in their own special **Helpful Hint** areas:

▷▷▷ **Helpful Hint:** Be sure to consult these for useful tips to use when solving thermodynamics homework problems and exercises—tips that you may or may not find in your primary textbook.

Material that (for the most part) can be read independently of the main body of text is sectioned off into gray-boxed areas that we refer to simply as *boxes*. These include examples, derivations, highlights, mathematical discussion, etc. The first box in this book appears on p. 13, with the title, **EXAMPLES: Law vs. Model**. Note that the term "box" refers only to *gray* boxes, and never to the special icon areas discussed above.

though technically, these also have a box around them...

One of the most confusing aspects of thermodynamics study is that many of the equations that one encounters apply only in certain situations. Such *conditional equations* are presented in this book using special formatting. Consider the example below:

February has 29 days. [leap year] (1.2)

The square brackets to the right of the equation indicate the conditions under which it applies.

▷▷▷ ***Don't* Try It!!** Don't bother trying to memorize all of the conditional equations in this book. This is useless unless you also memorize all of the conditions—and anyway, there are simply far too many of them!

⊳⊳⊳ **Helpful Hint:** In contrast, you *should* memorize the *definitions* of thermodynamic quantities—which according to Definition 1.1, must hold true *always*. You can then derive specific conditional results directly from the definitions, as needed.

In this book, quantities are underlined when defined.

We conclude this Introduction with a discussion of two "catch-phrases" that you will see repeated from time to time throughout this book, describing two principles that are useful to keep in mind.

⊳⊳⊳ **Science Doesn't Care.** This means that the natural world does not care about how easy or hard it is for us to understand it; it does what it does, regardless. One has to understand science on *its* terms, and not the other way around. Often, this means undermining our own preconceived notions about how things ought to work. On the other hand...

⊳⊳⊳ **It's OK to be Lazy.** You may have been taught that preconceived notions are a bad thing, but at least they provide a starting point. Rather than completely reinvent the wheel at every turn, scientists *often* borrow ideas from one area to apply to another. And why not, provided that it works? (though be wary of when it *doesn't* work...) It is important to realize that such a practice is not "cheating," and should not be considered shameful.

Isn't that a preconceived notion of its own?

The practice of borrowing terminology from everyday speech is one such example.

Note that we are *not* talking about the kind of laziness that might lead one to wait until the last possible moment before starting homework assignments or exam study... As in most fields, success in thermodynamics usually follows from steady, consistent, and disciplined effort.

Economy of thought and action is a virtue, not a vice, and well worth trying to master early in life.

Part I

Equilibrium

"In our wildest aberrations we dream of an equilibrium we have left behind and which we naively expect to find at the end of our errors."
—Albert Camus

"[Equilibrium] is a place where nothing ever happens."
—variation on "Heaven" by the *Talking Heads*

A Conceptual Guide to Thermodynamics, First Edition. Bill Poirier.

Companion website: http://www.conceptualthermo.com

Chapter 2

Philosophy of Thermodynamics

Anderson: Chap. 1, Sec. 2.5.1, Sec. 5.1, Sec. 5.2; **Atkins:** Sec. F.3, Sec. F.5, Sec. F.7; **Atkins-life:** Prolog; **Baierlein:** p. 13, Sec. 1.6, Sec. 1.7; **Callen:** pp. 2–3, Sec. 1-1, Sec. 1-2; **Cengel:** p. 1, Sec. 1–1, Sec. 1–4; **Chang:** p. xiii, Sec. 1.1, Sec. 1.3; **Elliott:** pp. 3–5; **Engel:** Sec. 1.1, p. 825; **Faure:** p. 155; **Kittel:** pp. 1–4, p. 7, p. 22, pp. 29–31, pp. 48–49; **Levine:** pp. 1–3; **McQuarrie:** pp. 764–766; **Moran:** p. v, p. 1, Sec. 1.1, Sec. 1.3.1; **Prausnitz:** pp. 1–2, Sec. 1.2, pp. 9–10; **Reif:** pp. vii–ix, pp. 1–4, p. 87, p. 152; **Sandler:** pp. 1–5; **Schroeder:** Preface; **Silbey:** pp. ix–xi, pp. 568–569; **Smith:** pp. xvii–xviii, p. 1, p. 647; **Tinoco:** p. 1, p. 13

2.1 Thermodynamics

Thermodynamics is the set of laws that govern the properties of matter at the macroscopic scale. As such, there is a direct link between thermodynamics and our own daily experience. Indeed, we human beings are equipped with sensory apparatus designed to directly measure two of the most important thermodynamic quantities—namely, *temperature* and *pressure*. So this gives us a deeply ingrained notion of what these quantities are, or at least how they should behave.

This, however, is a mixed blessing, because from a scientific perspective, there are other causes at play—less tangible and more abstract, and with a molecular origin—that govern how temperature and pressure behave, and what they really are. Thus, whereas our sensory intuition can in some cases be "right," in other cases, it can be wrong or misleading, and thereby lead us astray. Ultimately, mastery of thermodynamics—and science in general—requires looking at concepts beyond those of direct sensory experience.

Recall the discussion in Sec. 1.3.

Historically, thermodynamics was born at a time when the molecular nature of matter was still in doubt. Because thermodynamics is a law rather than a model (see Section 2.2), it is in fact possible to formulate a number of its principles purely at the macroscopic scale. This is called

A Conceptual Guide to Thermodynamics, First Edition. Bill Poirier.

Companion website: http://www.conceptualthermo.com

macroscopic or *classical* thermodynamics. On the other hand, we now know that the Laws of Thermodynamics ultimately arise from a *molecular* origin (see Section 2.3), and it is here that we must turn for a deeper, more rigorous understanding.

Think *Avogadro's number*, or $N_A \approx 6.02 \times 10^{23}$, individual molecules.

Molecularly speaking, macroscopic systems are *extremely* large and complex. Much of the utility of thermodynamics stems from the fact that it reduces all of this enormous molecular complexity down to just a few macroscopic parameters, known as *thermodynamic variables* (see Section 3.1). The net result is a greatly simplified description. In effect, the thermodynamic philosophy is to worry only about the "big picture," and not about what each individual molecule is doing.

Remember this when you are struggling with homework assignments; thermodynamics actually makes your life *much* easier!

Much of the richness of—and interest in—thermodynamics lies in reconciling the macroscopic and molecular viewpoints. For example, it is interesting to contemplate just where exactly it is that inherently macroscopic thermodynamic quantities come from, from a molecular standpoint. Temperature and pressure, for instance, are perfectly well-defined quantities for bulk matter, yet it is impossible to assign values for these quantities to *individual* molecules, nor even to a small number of molecules. For this reason, such quantities are sometimes called "emergent properties," meaning that precise values only "emerge" when the number of molecules becomes very large.

especially as this subject relates to materials science and engineering...

From the Texts: Some authors prefer "properties" over "quantities."

exactly *how* large is a key question—addressed, e.g., in cluster science...

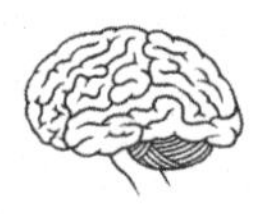

▷▷▷ **To Ponder...** Would scientists have ever even conceived of "temperature" and "pressure," had they not been able to sense these quantities directly? Might there exist other emergent properties, unknown to us by virtue of our lack of direct experiential faculties?

Such musings—though *not* what this book is really about—are difficult to avoid entirely in a conceptual treatment of thermodynamics. Rather, we seek primarily to provide a conceptual framework inasmuch as this serves to promote practical mastery of the subject across all disciplines. (Consult Chapter 17 and the website for a discussion of discipline-specific material.)

On the Website: http://www.conceptualthermo.com

▷▷▷ **Helpful Hint:** Try to understand the concepts clearly from the start—even if you do not see the point at first, and/or the relation to your chosen field. Have patience; it will connect eventually!

2.2 Scientific Models & Laws

In science and engineering, it is important to distinguish *models* from *laws*. Models are simplified, empirical descriptions of reality, built up over time in specific disciplines, based on observation of many specific examples. A scientific model is like a model car or airplane—lacking much of the detail and functionality of the original, and thus, not an exact description. In contrast, laws are *always* exact and true, because they derive from the fundamental principles of physics. Both models and laws are important, but they play different roles, and should be interpreted differently.

but "always" is of course a very long time, so one must be a bit cautious...

EXAMPLES: Law vs. Model

Law	**Model**
energy conservation	ideal gas "law"
Newton's Laws of motion	Lewis model of chemical bonding
quantum mechanics	Bohr model of the atom

Empirical science typically works something like the cycle presented in the box below. Step 3 happens for the same reason that courts do not always convict properly, and pundits do not always predict the future correctly—because models are based on empirical evidence, rather than fundamental principles. In any case, this can lead to a never-ending cycle, wherein the model is perpetually improved, but never to the point of completely evading all doubt and/or controversy.

CYCLE OF EMPIRICAL SCIENCE:

1. Some behavior is observed experimentally, for some class of systems.
2. A heuristic rule or model is formed.
3. The model is found to be not always correct.
4. The model is extended, for greater accuracy and/or applicability.
5. Go to Step 3.

Interjecting the laws of physics into the above cycle can help "cut to the chase" and resolve ambiguity. In contrast to models, new laws do not come around very often, and they are effectively irrefutable. On the other hand, laws also have their drawbacks. For example, they can be extraordinarily difficult to apply to real-world systems. Imagine, for instance, trying to simultaneously solve Newton's equations of motion for Avogadro's number's worth of particles!

for our purposes, anyway...

COMPARISON: Law vs. Model	
Law	**Model**
exact	approximate
complete	incomplete (why we keep inventing)
universal (applies to everything)	limited (applies to some things only)
complicated in practice	simple in practice

Are the principles of thermodynamics "laws" or "models"? Thermodynamics turns out to be a very, *very* special case, in that it provides the best of both worlds:

- It is an *exact law* of science, powerful and universal, yet...
- It is extremely *simple*, like a model.

The reason it can get away with this can be summarized in two words: *statistical mechanics*.

2.3 Statistical Mechanics

Thermodynamics is a law because it stems from a deeper theory called *statistical mechanics* that itself derives from the fundamental laws of physics. Specifically, statistical mechanics relates the macroscopic laws of thermodynamics to the physical laws governing the motion of individual molecules. The latter laws correspond to the "mechanics" part of statistical mechanics. This is taken to be either "classical" mechanics, for which Newton's Laws are used to describe the molecular motion, or "quantum mechanics," a more accurate and advanced treatment.

▷▷▷***Don't*** **Try It !!** Don't confuse classical *thermodynamics* with classical *mechanics*; the former is purely macroscopic, whereas the latter implies a molecular treatment.

The "statistical" part of statistical mechanics is equally important. This is the notion that for very large systems, individual molecules have very little impact on the macroscopic behavior—the "big picture" philosophy, alluded to in Section 2.1. In order to relate the molecular and macroscopic viewpoints, statistical mechanics relies on *statistical averages*, which for our purposes belong to one of two types. The first type is an average over all molecules or particles in the system—or over the *states* available to those

Hence, it took a very long time before the existence of molecules was confirmed.

molecules (see Section 3.3). The second type is an average over time, possibly of a single molecule (or even a single molecular *coordinate*; see Sections 5.2 and 5.4).

Although thermodynamics "rests on top" of statistical mechanics, the two can be *almost* completely separated—rather like "mind" and "brain." The idea of a total separation—essentially the purely macroscopic viewpoint—is undoubtedly appealing. However, a true conceptual understanding requires at least some delving down to the molecular level. We therefore adopt a "middle-of-the-road" approach, wherein a greatly simplified statistical mechanics is employed to define a few core thermodynamic quantities such as temperature and entropy. As always, the main goal is conceptual clarity, rather than mathematical rigor. In any event, in our treatment, the two "lobes" of thermodynamics and statistical mechanics remain fused into a single whole.

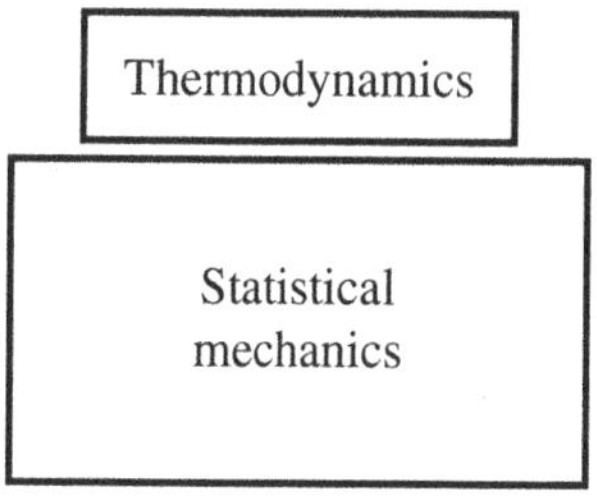

Albert Einstein once remarked:

> *[Statistical thermodynamics] is the only physical theory of universal content which I am convinced will never be overthrown...*

Why might he have said this? Perhaps because, despite being a mathematically very advanced topic, conceptually, statistical mechanics is really about nothing more than just counting—and the Law of Counting will never change.

Thermodynamics

Statistical mechanics

Conceptual relationship between thermodynamics and statistical mechanics. Thermodynamics is just the "tip" of the statistical mechanics "iceberg."

who developed more than his fair share of new physical theories...

Chapter 3

Thermodynamic States, Variables & Quantities

Anderson: Sec. 2.1, Sec. 2.2, Sec. 2.4, Sec. 3.1; **Atkins:** Sec. F.3, Sec. F.5, Sec. F.7, Sec. 1.1; **Atkins-life:** Sec. F.2; **Baierlein:** p. 13, Sec. 1.6; **Callen:** Sec. 1-3, Sec. 1-9, Sec. 12-2, Sec. 21-9; **Cengel:** p. 1, Sec. 1–1, Sec. 1–2, Sec. 1–6, Sec. 3–1; **Chang:** Sec. 1.2, Sec. 1.3; **Elliott:** Sec. 1.2, Sec. 1.4; **Engel:** Sec. 1.2; **Faure:** Sec. 11.1; **Kittel:** Chap. 1, p. 29; **Levine:** Sec. 1.2, Sec. 1.4; **McQuarrie:** p. 769; **Moran:** Sec. 1.2, Sec. 1.3, Sec. 1.4; **Prausnitz:** Sec. 1.1, Sec. 2.1; **Reif:** pp. 47–52; **Sandler:** Sec. 1.1, Sec. 1.2; **Schroeder:** Sec. 1.1, pp. 163–164; **Silbey:** Sec. 1.1; **Smith:** Sec. 1.2, Sec. 1.3; **Tinoco:** pp. 26–30

3.1 Thermodynamic Variables & Quantities

As discussed in Chapter 2, thermodynamics is able to reduce all of the molecular-scale complexity of macroscopic systems down to just a few thermodynamic variables. These are:

- temperature, T
- pressure, P
- volume, V
- number of particles, N

Remarkably, these four variables alone determine *everything there is to know* about the system at the macroscopic scale. In other words, from knowing the values of the four thermodynamic variables, the value of *any* other macroscopic quantity may be predicted with perfect precision, without the need to conduct an actual laboratory measurement. Although there are some caveats, this is an amazing fact—and the reason that we can regard thermodynamics as a being a true *law*, rather than a mere model. More technically, we say that thermodynamics is *macroscopically complete*.

If you learn nothing else from this chapter, let it be this.

at least in principle…

Recall from Sec. 2.2 that completeness is the hallmark of a *law*, rather than a model.

A Conceptual Guide to Thermodynamics, First Edition. Bill Poirier.

Companion website: http://www.conceptualthermo.com

On to the caveats. First, in addition to the four thermodynamic variables, knowledge of the particular substance of which the system is composed is also required, in order to make specific macroscopic predictions. Second, thermodynamics does *not* tell us everything there is to know about the individual molecules that make up the system. As a predictive theory, thermodynamics is decidedly *incomplete* at the molecular scale. Still, molecular information is not actually necessary, in order to understand with perfect precision what is happening at the macroscopic scale.

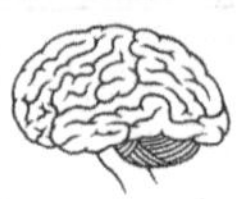

▷▷▷ **To Ponder...** In fact *greater* macroscopic precision actually requires there to be *more* molecules in the system—and thus *more* unknown molecular information. This is one of many counterintuitive results that has been proven using statistical mechanics...

Next, we clarify some terminology. By *thermodynamic system*, we simply mean "a macroscopic amount of stuff" (i.e., bulk matter). *Macroscopic*, in turn, means that the number of molecules or particles is on the order of *Avogadro's number,*

Molecules vs. particles: The latter is a bit more general and "physicsy," but we treat these as synonyms.

$$N_A \approx 6.02214 \times 10^{23}. \qquad \text{[Avogadro's number]} \quad (3.1)$$

A *thermodynamic quantity* is any macroscopic property of the system that can be measured as a number; this includes the four thermodynamic variables, among many others. To specify a quantity with *perfect precision* means "to any desired accuracy"—i.e., to any number of significant digits. We also take a *substance* (or *pure substance*) to be a thermodynamic system whose constituent particles are all of the same kind. This assumption is presumed throughout this book.

Equivalently, we can say that the quantity is "well defined."

although special chapters addressing phase transitions, mixtures, and chemical reactions are available on the website (see Chap. 17)...

For a pure substance, the most fundamental thermodynamic quantity imaginable is the *amount of substance*—i.e., how much "stuff" there is. There are two ways to reckon this. The molecular-scale quantity is N, the number of particles (molecules). There is also a macroscopic measure of the amount of substance, the so-called *number of moles*, n, defined as

$$n = \frac{N}{N_A}. \qquad \text{[number of moles]} \quad (3.2)$$

There is no fundamental difference between N and n; the distinction is merely one of scale. In statistical mechanics and physics, N is usually preferred, whereas in thermodynamics and chemistry, n is typically used. Engineers eschew both quantities, preferring to work with the *system mass*, $Nm = nM$, where m is the mass per particle, and M is the mass per mole (molar mass). In any case, we have now defined our first thermodynamic quantity, and we did so by following Einstein's advice—i.e., by *counting*.

3.2 More on Thermodynamic Quantities

There are many thermodynamic quantities that can be defined or measured. We usually denote these quantities using capital letters—with 'X' representing a generic choice.

and with perfect precision, no less, provided that N is sufficiently large...

Thermodynamic quantities can be usefully divided into two basic types:

extensive vary proportionally with n.
intensive independent of n.

EXAMPLES: Extensive vs. Intensive Quantities

Extensive	**Intensive**
amount of substance, n or N	temperature, T
system mass, $Nm = nM$	pressure, P
volume, V	molar volume, $V_m = V/n$
internal energy, U	chemical potential, μ

Note that any extensive quantity can be converted into an intensive quantity, simply by dividing by n. The resultant *molar* quantity receives an 'm' subscript.

One exception is 'M', the molar mass.

Any quantity in science can be characterized with both a number and a unit. Since thermodynamics describes macroscopic systems, we generally use macroscopic units—almost always *Système International* (SI) units. Note that thermodynamic quantities always describe the *whole system*, rather than individual particles. This is true even for molecular-scale thermodynamic quantities, such as the per-particle average, $\langle X/N \rangle$.

the particle analog of the *molar* average, X_m...

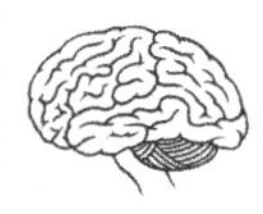

▷▷▷ **To Ponder...** For a particular city, it may well be the case that the average number of children per family is 2.3. But there is not a *single* family in that city that actually has 2.3 children!

Although SI units are very standard—meters (m) for length, Joules (J) for energy, Pascals (Pa) for pressure, etc.—it is still possible to get into trouble. For students of thermodynamics, the most common error *by far* is to take the liter (L) as the SI unit of volume.

▷▷▷***Don't* Try It!!** Don't *ever* take the liter (L) to be the SI unit of volume! This is the *single most common* error made by students of thermodynamics—and it appears very often on multiple choice exams.

The correct SI unit of volume is the *cubic meter* (m^3),which is exactly equal to 1000 liters. Students often make the "liter/cubic meter" error when computing energy as pressure times volume. In SI units, we have

$$1\ \text{J} = 1\ \text{Pa m}^3, \qquad \textit{not} \quad 1\ \text{J} = 1\ \text{Pa L}$$

▷▷▷**Helpful Hint:** Because the "liter/cubic meter" error is *so* common, when solving problems, you should always be on the lookout for results that are either 1000 times larger or smaller than they should be.

Another potential source of trouble is the use of the Celsius or centigrade scale for temperature. In thermodynamics, the Kelvin temperature scale should almost always be used.

Other *absolute temperature scales* (Sec. 13.2) are also sometimes used—e.g., Rankine (R) and fundamental (Sec. 10.4).

▷▷▷***Don't* Try It!!** Don't *ever* use Celsius or centigrade in thermodynamics! If a problem gives you the temperature in degrees Celsius (°C), convert to kelvin (K) before doing anything else.

3.3 Thermodynamic & Molecular States

When a measurement or scientific theory provides a complete description of a physical system, we say that the system's *state* is determined. Conversely, knowledge of the system state suffices to predict the value of any desired quantity. From Section 3.1, we learn that the four thermodynamic variables (T, P, V, N) collectively specify the macroscopic or *thermodynamic state* of a thermodynamic system. Insofar as thermodynamics is concerned, all systems of a given molecular type, with the same values for these four variables, occupy the same thermodynamic state. Since these systems must exhibit the same value for *any* thermodynamic quantity, there is simply no macroscopic way to distinguish them.

According to the laws of physics, however, the system can also be described in terms of its *molecular state*. For a single molecule, this state is determined from the following:

- particle *position*, (x, y, z)
- particle *velocity*, (v_x, v_y, v_z)
- internal structure

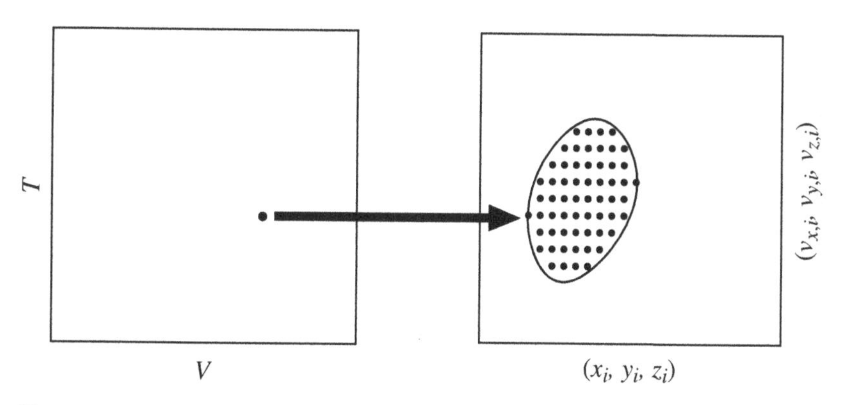

Figure 3.1 One-to-many relationship between thermodynamic states and molecular states. For a given thermodynamic state of the system [the point (T, V) in the left plot; don't worry about P and N], there are many possible molecular states (the set of points in the right plot), amongst which thermodynamics cannot distinguish.

Although real molecules possess internal structure, in thermodynamics, we often treat molecules as if they were "point particles" (see Section 5.3)—in which case, only the position and velocity are required to specify the particle's state.

For an entire macroscopic system, the molecular state is defined to be the collection of single-molecule states for the constituent molecules, taken together. Specifying the molecular state of the whole system thus requires at least $6N$ coordinates—the three position components, (x_i, y_i, z_i), plus the three velocity components, $(v_{x,i}, v_{y,i}, v_{z,i})$, for each molecule, i, where the label i is an integer with $1 \leq i \leq N$.

▷▷▷ ***Don't* Try It!!** Don't confuse the state of a *single molecule* with the molecular state of the *whole system*. Only when the system consists of just one molecule—i.e., when $N = 1$—are the two concepts the same.

As discussed, knowing the thermodynamic state of a system does not provide specific information about individual molecules. Since most of these molecular details remain macroscopically hidden, there is a very large number of possible molecular states that correspond to any given thermodynamic state—i.e., that are consistent with what *is* known macroscopically about the system. This one-to-many relationship between thermodynamic and molecular states, indicated in Figure 3.1, is extremely important.

In Chap. 10, we learn that this "missing" molecular information gives rise to the *entropy*.

Chapter 4

Zeroth Law & Thermodynamic Equilibrium

Anderson: Sec. 2.3, Sec. 2.5.1, Sec 13.2, Sec. 13.4; **Atkins:** Sec. 1.1, Sec. 1.2; **Atkins-life:** Sec. F.2(c), Sec. 1.1; **Baierlein:** Sec. 1.1, Sec. 1.2; **Callen:** Sec. 1-5, Sec. 1-6, Sec. 2-1, Sec. 2-2, Sec. 2-4, Sec. 2-7; **Cengel:** p. 1, Sec. 1–3, Sec. 1–4, Sec. 1–6, Sec. 1–8, p. 123; **Chang:** Sec. 1.2, Sec. 2.1, Sec. 2.2, Sec. 2.3; **Elliott:** p. 16, pp. 18–22, Sec. 1.5; **Engel:** Sec. 1.3, Sec. 1.4; **Faure:** p. 156; **Kittel:** pp. 1–3, pp. 39–41, p. 48, Chap. 6; **Levine:** pp. 4–5, Sec. 1.3, Sec. 1.4, Sec. 1.5, Sec. 1.7, Sec. 2.8, Sec. 8.1; **McQuarrie:** Sec. 16–1, Sec. 16–2; **Moran:** Sec. 1.3.4, Sec. 1.7, Sec. 11.1; **Prausnitz:** p. 2, Sec. 2.1, pp. 31–32, p. 35; **Reif:** Sec. 2 · 6, Sec. 2 · 7, Sec. 2 · 8, Sec. 3 · 4, Sec. 3 · 9, Sec. 5 · 1; **Sandler:** p. 3, Sec. 1.1, Sec. 1.3; **Schroeder:** Sec. 1.1, Sec. 1.2; **Silbey:** Sec. 1.2, Sec. 1.3, Sec. 1.4; **Smith:** Sec. 1.4, Sec. 1.5, Sec. 1.6, Sec. 2.6; **Tinoco:** pp. 28–30

4.1 Equation of State

Although there are four thermodynamic variables in all (Section 3.1), under most circumstances, this set can be reduced down to just *two* independent variables. Throughout most of this book (until Chapter 15), the thermodynamic system of study is presumed to be *closed*, meaning that no particles can flow in or out. Consequently, N (or n) is a constant, rather than a variable. This reduces the set of thermodynamic variables down to three—i.e., T, P, and V.

Furthermore, for a system in *thermodynamic equilibrium* (Section 4.2), it has been established that only *two* of these three variables are independent. The third (dependent) variable is uniquely determined by the two independent variables, from which it can be obtained using the *equation of state*. This is a single equation involving all three variables, (T, P, V), that constrains the allowed values of those variables—thereby specifying which thermodynamic states are in equilibrium.

established both experimentally and via statistical mechanics…

The words "in equilibrium" are very important! We will explain exactly what this means in Sec. 4.2.

A Conceptual Guide to Thermodynamics, First Edition. Bill Poirier.

Companion website: http://www.conceptualthermo.com

EXAMPLE: Splitting a Pizza Among Three Friends

Tom, Paul, and Valerie are splitting a pizza with eight slices total. Let 'T', 'P', and 'V' denote the number of slices eaten, respectively, by each of these three friends. In principle, each feaster could wind up eating anywhere from zero to eight slices. However, T, P, and V are not all independent; assuming that the friends completely devour the pizza (a reasonable assumption!), then the total number of slices eaten must be equal to eight. Thus, there is a *constraint* on T, P, and V, which can be written as a single, simple equation:

$$T + P + V = 8 \tag{4.1}$$

Equation (4.1) can be thought of as the "equation of state" for the pizza "system." Because of this constraint, there are really only *two* independent variables needed to describe this system completely—i.e., to specify exactly how many slices everyone ate. We can use the equation of state to specify any one of the three variables as a function of the other two.

For example, if we know T and V, we can get P by rearranging Equation (4.1) as follows:

$$P(T, V) = 8 - T - V \tag{4.2}$$

In Equation (4.2), T and V are being treated as the independent variables, and P as the dependent variable. Note, however, that we could just as easily treat P and V as the independent variables (for instance), which would lead to an equation for T:

$$T(P, V) = 8 - P - V \tag{4.3}$$

All of the conclusions reached here regarding the pizza "system" have their analog in the thermodynamics world. In particular, every pure substance is described by some equation of state that specifies which thermodynamic states are in equilibrium. The precise form of the equation of state depends on the particular substance—thereby accounting for many of the specific material properties of that substance.

Two important points should be kept in mind:

From the Texts: Some authors may not stress these two points sufficiently.

1. The equation of state refers *only* to systems in thermodynamic equilibrium. (The out-of-equilibrium case is discussed on pp. 26–27.)
2. The equation of state does not inherently treat any *one* thermodynamic variable differently than the other two.

Point 2 is particularly key, because it means that the choice of independent variables is flexible; it can be changed as desired, from one homework problem or laboratory experiment to the next. Thus, one might explicitly control T and V in one experiment, allowing the system to find its own P

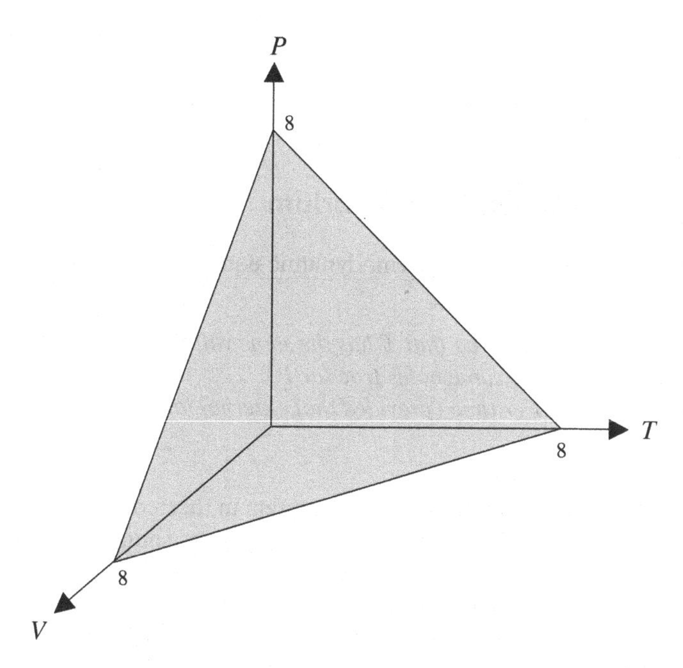

Figure 4.1 **Pizza "system" equation of state.** Plot of pizza "system" equation of state, as a surface in (T, P, V) space. The vertical axis, P, is being treated as the dependent variable, as per Eq. (4.2). We can also obtain the $T(P, V)$ surface of Eq. (4.3), by simply rotating the figure until the T axis is vertical. The surface is still really the same; the new orientation simply reflects the new choice of independent variables.

as it achieves equilibrium; in another experiment, T and P might be fixed, in which case the system automatically expands or contracts until the equilibrium V value is reached.

▷▷▷ **Helpful Hint:** You have the freedom—*and also the obligation*—to choose which two of (T, P, V) you will take to be your independent variables. Get in the habit of consciously making this choice, at the start of each new problem that you encounter. Choose wisely! Once the decision is made, you must stick with it until the problem is completed.

By following the above Helpful Hint, you may avoid much of the confusion that often saddles beginning students of thermodynamics. A good choice of independent variables can also save you a lot of work. Remember, **It's OK to be Lazy**.

4.2 Thermodynamic Equilibrium

Definition 4.1 *A system in* thermodynamic equilibrium *is one for which* (T, P, V):

1. *are well defined (meaning that* T *has the same value throughout the whole system, and that the same holds true for* P*).*
2. *remain constant over time (provided that external factors do not change).*

Since the thermodynamic *variables* for a system in thermodynamic equilibrium are well defined and constant over time, the same must also be true of the thermodynamic *state* of that system.

This does not mean that the *molecular* state is constant over time—which would imply that the individual molecules are "frozen in place." This is far from the case; at the molecular scale, particles are constantly moving, colliding, etc. But when the system is in thermodynamic equilibrium, all of these individual molecular changes tend to cancel each other out, statistically speaking, so that there is no net change at the macroscopic scale. Referring to Figure 3.1, one can imagine the system bouncing around from one point in the right plot (i.e., one molecular state) to another, but always within the region indicated. The corresponding thermodynamic state (the point indicated in the left plot) remains unchanged.

This is a nice example of the one-to-many relationship between thermodynamic and molecular states.

One can also imagine a different set of circumstances, under which the molecular state might change so much that it ends up moving *outside* of the Figure 3.1 (right plot) region. In this case, the final molecular state corresponds to a new *thermodynamic* state as well (i.e., a new point in the left plot). Such a macroscopic change of state can only occur as the result of the system somehow getting out of equilibrium.

EXAMPLE: Inflated Tire System

At some initial time, an inflated tire has a certain set of (T, P, V) values that lie on the equation of state.

Question: *Is the inflated tire system in equilibrium?*

Yes. The tire will remain in the same thermodynamic state indefinitely (meaning that T, P, and V will not change over time), provided external factors do not change.

EXAMPLE: Punctured Tire System

Now, consider what happens when the tire is punctured, allowing the enclosed air to rush out.

Question: *Is the punctured tire system still in equilibrium?*
No, because a *sudden external change* has been introduced, resulting in P values that vary from one point to another within the tire, and also over time. Ultimately, though, the resultant macroscopic changes give rise to a new equilibrium thermodynamic state—a flat tire!

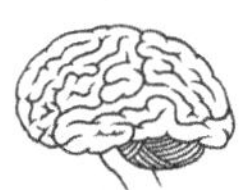

⊳⊳⊳ **To Ponder...** To say that the air "rushes" out of the punctured tire is actually misleading. Most of the air molecules do indeed wander out of the tire—but moving no faster than prior to the puncture. Moreover, there are also a great many air molecules that wander *into* the tire from the outside—perhaps only a few less, in relative terms, than the number that wander out. This difference is statistically significant, however, giving rise to a macroscopic change that we perceive as *wind.*

Thus, when the macroscopic notion of "wind velocity" is analyzed at the molecular scale, it is actually found to have more to do with the *number* of air molecules moving in a preferred direction, than with their speeds. It would therefore be more accurate to refer to a "big" wind than to a "strong" or "fast" wind.

The word "typhoon" is derived from the Chinese and Japanese for "big wind."

4.3 Zeroth Law

By virtue of Definition 4.1 (p. 26), together with the definition of the equation of state (p. 23), there are two ways in which a system can be *out* of equilibrium:

1. (T, P, V) are well defined (same throughout system), but do not lie on the equation of state.
2. T and/or P vary from one point to another within the system.

Whenever a system gets out of equilibrium—however this has occurred—it undergoes a spontaneous macroscopic change, until a new equilibrium state is achieved.

Situation 2 above is often observed when the system consists of two *subsystems*—each initially in an equilibrium state, but not the *same* equilibrium state (i.e., different T and/or P values). When the two subsystems,

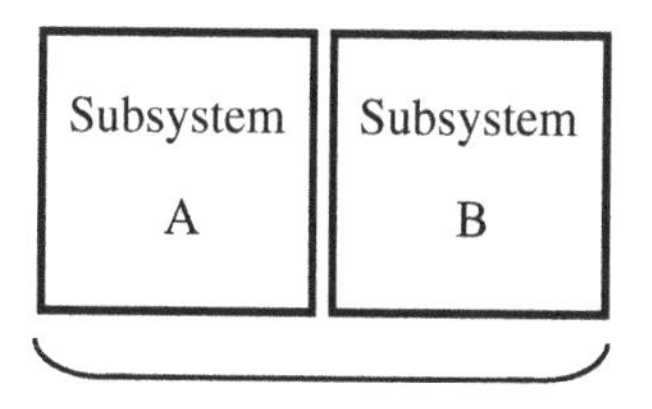

Two subsystems, A and B, brought together to form a single combined system.

A and B, are suddenly brought into contact and allowed to interact, they form one big system that is no longer in equilibrium. A spontaneous change then ensues, until the combined system achieves equilibrium. At this point, both subsystems have the same T and P throughout—but these values are different than what either subsystem started with.

Now imagine that in the above subsystems example, A and B *start out* with the same T and P values. Now when they are brought together, nothing happens; the combined system is already in thermodynamic equilibrium, because T and P values are the same throughout. So in this case, there is no macroscopic change, and we say that A and B are *in equilibrium with each other*. Treating T and P as the independent variables, we also see that A and B are in the same thermodynamic state. This leads to the

Zeroth Law: *If A is in equilibrium with B, and B is in equilibrium with C, then A is in equilibrium with C.*

What is the real significance of this law? It simply confirms the existence of thermodynamic states, and the macroscopic completeness of thermodynamics.

It also establishes exactly why thermometers work.

We have discussed the idea of bringing subsystems into "contact," but have not yet specified how this is done. It turns out that there are two kinds of contact, each associated with a different kind of equilibrium, and a different intensive thermodynamic variable.

actually more than two, but only two for closed systems comprised of a pure substance in a single phase...

Mechanical contact is associated with P, and with *mechanical equilibrium.* To bring two subsystems into mechanical contact, the wall that divides them must be allowed to move. Imagine that there is gas on either side of the dividing wall, initially at two different pressures, $P_A > P_B$. The pressure difference exerts a macroscopic *force*, F, tending to push the wall towards B, the low-pressure side:

but for closed systems, the movable wall must still not allow particles through...

$$F = (P_A - P_B) \times \langle \text{area of wall} \rangle. \tag{4.4}$$

Thus, if A and B are brought into mechanical contact, so that the fixed wall is suddenly allowed to move, it will do so—expanding subsystem A, compressing subsystem B, and transferring energy from A to B in the form of *work*—until finally, $P_A = P_B = P$, and mechanical equilibrium is restored.

From the Texts: There are some excellent explanations and figures to be found in the reference textbooks, notably: **Atkins**, Sec. 1.1; **Engel**, Sec. 1.3; **Sandler**, Sec. 1.4; **Silbey**, Sec. 1.1, Sec. 1.2.

Thermal contact is associated with T, and with *thermal equilibrium.* Here, the dividing wall must be *diathermic*, or thermally conductive, in order for thermodynamic change to occur. If two subsystems are suddenly brought into thermal contact, and there is a nonzero temperature difference $[(T_A - T_B) > 0]$, then energy will flow in the form of *heat* from the hotter subsystem (A) to the cooler one (B)—until $T_A = T_B = T$, and thermal equilibrium is restored.

Most real-world systems consist of more than one substance and/or phase of matter (e.g., gas, liquid, solid). In such cases—and also for *open* systems (the opposite of "closed")—another form of contact known as *diffusive contact* also plays a key role. Analogs exist in chemical reactions, electrochemistry, and various other applications. Diffusive contact is discussed in Chapter 15; applications for which it is relevant are discussed in Chapter 17 and on the website.

On the Website:
http://www.conceptualthermo.com

4.4 Ideal Gases & Non-ideal Systems

No thermodynamics book would be complete without at least a mention of the *ideal gas law*,

$$PV = nRT = NkT. \qquad \text{[ideal gas]} \quad (4.5)$$

However, since it is no doubt already covered extensively in your primary textbook, along with non-ideal systems, we limit discussion here to just a few highlights.

HIGHLIGHTS: The Ideal Gas Law

- It is a single equation in all three thermodynamic variables, (T, P, V), and thus an *equation of state*.
- It predicts the same molar volume of $V_m = 22.4$ L/mol for *all* substances at standard T and P ($T^\circ = 273.15$ K, $P^\circ = 1$ atm = 101325 Pa).
- It assumes that the system particles are *independent* or *noninteracting* (Section 5.3).
- It is not actually a law, but a *model*, valid only in the limit, $V_m \to \infty$.

Note that the ideal gas equation of state is the same for all substances. In reality—i.e., for non-ideal gases and condensed phases (liquids and solids)—the actual equation of state varies considerably from one substance to another. In general, the ideal gas law breaks down as $T \to 0$, because it predicts that the molar volume $V_m \to 0$ in this limit—i.e., that the system becomes vanishingly small as it approaches absolute zero temperature—which is physically incorrect.

What in fact happens when T becomes small is that the system undergoes a *phase transition* to a condensed phase. The latter has a very small—but nevertheless nonzero—V_m value, which is also highly substance-specific (as is the transition temperature). Consequently, whereas the ideal gas law is usually a good approximation for real gases, it is *completely* invalid for condensed phases.

▷▷▷ ***Don't* Try It !!** Don't *ever* use the ideal gas law for a liquid or a solid! This is the *second* most common error made by students, after the "liter/cubic meter" error of Section 3.2.

It is useful to characterize real substances in terms of the *compressibility factor*:

$$Z = \frac{PV}{nRT} \qquad \text{[compressibility factor]} \qquad (4.6)$$

For ideal gases, $Z = 1$. For other materials, the sign and magnitude of $(Z - 1)$ provide important information about the character and extent, respectively, of the non-ideality (see Sections 5.3, 9.3, and 15.3).

Much like n and N, the gas constant R and the Boltzmann constant k are macroscopic and molecular versions of the same quantity—being essentially, a conversion factor between temperature and energy (per mole for R; per particle for k). Thus,

$$R = kN_A, \quad \text{and} \quad nR = Nk, \qquad (4.7)$$

and all parts of Equation (4.5) have (macroscopic) dimensions of energy.

⊳⊳⊳**Helpful Hint:** Learn to convert between macroscopic and molecular scale expressions, such as nR and Nk, with ease. Unlike in Equation (4.5), we will generally present only one form or the other, in any given expression.

Part 

Energy

"If we wish to find in rational mechanics an a priori *foundation for the principles of thermodynamics, we must seek mechanical definitions of temperature and entropy."*

—J. Willard Gibbs

"Available energy is the main object at stake in the struggle for existence and the evolution of the world."

—Ludwig Boltzmann

A Conceptual Guide to Thermodynamics, First Edition. Bill Poirier.

Companion website: http://www.conceptualthermo.com

Chapter 5

Molecular Energy, Internal Energy, & Temperature

Anderson: Sec. 1.3, Sec. 3.1.1, Sec. 3.2; **Atkins:** Sec. F.4, Sec. F.5, pp. 21–22, Sec. 2.2; **Atkins-life:** Sec. F.3, Sec. 1.2(c), Sec. 1.5; **Baierlein:** Sec. 1.1, p. 7, Sec. 1.3, Sec. 13.3; **Callen:** Sec. 1-4, Sec. 2-4, Sec. 16-11; **Cengel:** Sec. 1–8, Sec. 2–1, Sec. 2–2; **Chang:** Sec. 2.2, Sec. 2.6, Sec. 3.2, p. 530; **Elliott:** Sec. 1.2, Sec. 1.6, Sec. 2.12, Sec 7.2; **Engel:** Sec. 1.2, Sec. 2.2, Sec. 31.8; **Faure:** Sec. 11.2; **Kittel:** pp. 18–23, pp. 39–42, pp. 72–78, p. 164; **Levine:** Sec. 2.1, Sec. 2.4, Sec. 2.11, pp. 101–102, Sec. 14.2, Sec. 14.3; **McQuarrie:** Sec. 19–2, Sec. 19–3, Sec. 25–1; **Moran:** Sec. 1.7, Sec. 2.1, Sec. 2.3; **Prausnitz:** Sec. 2.1, Chap. 4; **Reif:** Sec. 3 · 5, Sec. 4 · 1, Sec. 4 · 3, Sec. 6 · 1, Sec. 6 · 2, Sec. 7 · 6; **Sandler:** Sec. 1.4, Sec. 1.5, Sec. 1.6; **Schroeder:** pp. 11–17, Sec. 2.4, Sec. 2.5; **Silbey:** Sec. 1.3, Sec. 2.2, Sec. 16.9; **Smith:** Sec. 1.5, Sec. 2.2, Sec. 2.5; **Tinoco:** pp. 21–24, pp. 46–47, p. 64, pp. 156–157

This chapter is about energy, the most important quantity in all of science. Our main goal is to define the *internal energy*—a new thermodynamic quantity, denoted 'U'. Though internal energy is macroscopic, the definition requires an understanding of energy at the molecular scale, which we accordingly also consider. Along the way, we will come to better understand how molecules interact, and what qualitative effect this interaction has on the macroscopic behavior of the system (e.g., on the equation of state).

We also define the *temperature*, denoted 'T'.

5.1 Energy at the Molecular Scale

From the laws of physics, we learn that there are two kinds of energy: *potential energy* and *kinetic energy*. At the molecular scale, it is convenient to imagine a third kind of energy possessed by individual molecules, which

A Conceptual Guide to Thermodynamics, First Edition. Bill Poirier.

Companion website: http://www.conceptualthermo.com

we call the *internal state energy*. Each type of energy is associated with a different type of molecular coordinate.

CORRESPONDENCE: Molecular Coordinates & Energy

Coordinate type	**Energy type**
position, (x, y, z)	potential energy, E_P
velocity, (v_x, v_y, v_z)	kinetic energy, E_K
internal structure	internal state energy, E_I

When the subscript i is applied to the quantities in the box above, this refers to the specific molecule i. Thus, $E_{K,i}$ is the kinetic energy associated with the translational (center-of-mass) motion of molecule i. "Potential energy" refers to the intermolecular interactions among molecules. Strictly speaking, E_P depends on the *relative* (to other molecules) molecular positions, rather than on the *absolute* (center-of-mass) molecular positions, (x_i, y_i, z_i). However, this distinction does not matter for our purposes. Finally, E_I represents the energy associated with *intramolecular* interactions. For most thermodynamics applications, $E_{I,i}$ can be regarded as the chemical energy stored in molecule i—if indeed, it need even be regarded at all.

and in many cases, also the relative *orientations*...

sometimes **It's OK to be Lazy**...

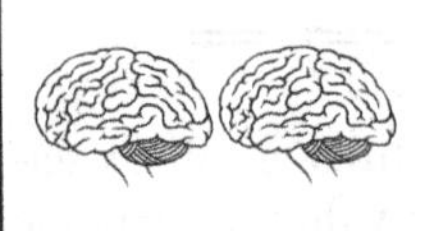

⊳⊳⊳ **To Ponder...*at a deeper level.*** In reality, E_I is comprised of intramolecular kinetic and potential energy, of various kinds. The internal structure of molecules can be quite complicated, requiring quantum mechanics for a complete understanding.

The energy of the molecular state of the whole system, denoted 'E', is obtained by summing the three kinds of energy over all molecules:

$$E = \sum_{i=1}^{N} \left(E_{P,i} + E_{K,i} + E_{I,i}\right) \tag{5.1}$$

According to the laws of physics (specifically conservation of energy), E remains constant over time, provided that the system is "isolated" (see Section 7.1). This is not the case for most thermodynamic systems, which are usually in thermal and/or mechanical contact with their surroundings. Consequently, energy can flow freely in and out of the system over time, and so E is not constant. This is true even when the system is in complete thermodynamic (i.e., both mechanical and thermal) equilibrium.

Over time, the molecular state changes, but the thermodynamic state does not (marginal note on p. 26).

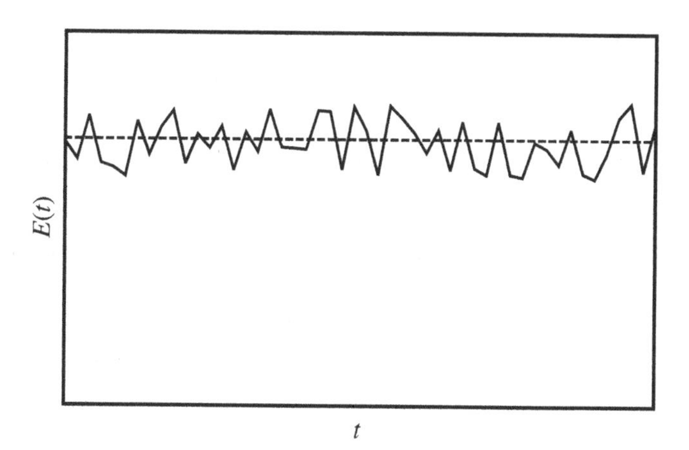

Figure 5.1 **Molecular state energy as a function of time.** Energy of the molecular state of the whole system, E, as a function of time, t (solid curve), for a system in thermodynamic equilibrium. As indicated, $E(t)$ itself is not constant, but tends to fluctuate around its time-averaged value, $\langle E(t) \rangle$—indicated by the horizontal dashed line. This constant value is (essentially) the *internal energy*, U.

5.2 Internal Energy

Because E can change over time even when the thermodynamic state does not, this means that E itself *cannot* be a thermodynamic quantity, according to Definition 4.1 (p. 26). So how *can* we define a meaningful thermodynamic energy quantity? This is where the idea of statistical averaging comes in—specifically, time averaging (Section 2.3).

A plot of the molecular state energy, E, as a function of time, t, for a macroscopic system in thermodynamic equilibrium, would look something like Figure 5.1. Note that $E(t)$ oscillates very quickly, but never gets far from the time-averaged value. These oscillations are called *fluctuations*. According to statistical mechanics, fluctuations quickly cancel out when averaged over macroscopic time scales, and can therefore be *ignored*—provided that the system is both large ($N \to \infty$), and in equilibrium.

In practice, the time-averaged quantity,

$$\langle E(t) \rangle = \frac{1}{t_{\text{final}}} \int_0^{t_{\text{final}}} E(t)\, dt, \tag{5.2}$$

is much more useful than E itself. Moreover, for a system in equilibrium, $\langle E(t) \rangle$ is constant over time, and may therefore be regarded as a true thermodynamic quantity. We take this quantity to be the *internal energy* (after subtracting the internal state energy).

Definition 5.1 *For a system in equilibrium, the* internal energy, *denoted 'U', is defined to be the time average of the molecular state potential-plus-kinetic energy:*

$$U = \left\langle \sum_{i=1}^{N} (E_{P,i} + E_{K,i}) \right\rangle = \sum_{i=1}^{N} \left(\langle E_{P,i} \rangle + \langle E_{K,i} \rangle \right). \quad \text{[equilibrium]} \quad (5.3)$$

Note from Equation (5.3) that U can be obtained from the time averages of the *individual molecule* potential and kinetic energies, $\langle E_{P,i} \rangle$ and $\langle E_{K,i} \rangle$. Note also the absence of $E_{I,i}$. The internal state energy is not included in the definition of U, because it plays no direct role in many "traditional" thermodynamic processes such as gas expansions.

though it obviously plays a critical role in chemical reactions (Chap. 17)...

⊳⊳⊳ ***Don't*** **Try It!!** Don't confuse *internal energy*, U, with *internal state energy*, E_I. A key difference is that U is a thermodynamic state quantity, and E_I is a molecular state quantity. Perversely, the internal state energy is the *only* type of molecular energy that does *not* contribute to the internal energy!

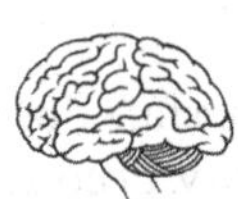

⊳⊳⊳ **To Ponder...** Why on earth, then, do they call U the "internal" energy? Wouldn't "thermodynamic energy" be a much better and less confusing term? Probably yes... In any case, the answer underscores the perennial clash between the molecular and macroscopic viewpoints. The energy U is "internal" to the thermodynamic system in a *macroscopic* sense—as opposed to being "internal" to the molecules (as is E_I).

From the Texts: A nice discussion may be found in **Cengel**, p. 55, and **Smith**, Sec. 2.2.

The internal energy includes, say, the energy available to do work, but does not include, say, the translational kinetic energy arising from movement of the whole thermodynamic system (i.e., the experimental apparatus) through space.

Ordinarily, we like our experiments to stay put...

⊳⊳⊳ ***Don't*** **Try It!!** Don't confuse a *molecular* quantity (e.g., E) with its statistical average (e.g., $\langle E \rangle$), which is very often a true *thermodynamic* quantity. In fact, statistical averaging is one of the most common—and direct—ways of constructing new thermodynamic quantities.

Sections 5.3 and 5.4, respectively, will discuss potential and kinetic energy in more detail.

5.3 Intermolecular Interactions & the Kinetic Model

As discussed in Section 5.2, internal state energy is ignored in the calculation of U. From a molecular viewpoint, this is equivalent to making the assumption of point particles. As per the discussion in Section 3.3, point particles have no internal structure, and therefore no internal state energy. Very often this is a good assumption, thermodynamically speaking, provided that the actual internal state does not change.

Note that some authors refer to point particle systems as "monoatomic" systems—e.g., argon gas. This is reasonable nomenclature, but only to a point; even monoatomic systems have internal *atomic* structure. Conversely, even *non*-monoatomic systems, such as diatomic nitrogen, can in some ways be treated as point particle systems. We therefore prefer the "point particle" terminology.

From the Texts

Another commonly made assumption in thermodynamics is the ideal gas assumption. We have already encountered this in Equation (4.5), the ideal gas equation of state. Our goal here is to discuss what this assumption means at the molecular scale. In fact, it means that there are no intermolecular interactions at all—i.e., that $E_{P,i} = 0$, and so all of the individual molecules are noninteracting, and behave completely independently. We thus also call them *free particles*.

The ideal gas is an extremely important special case, and the *only* one for which the statistical mechanics problem can be solved exactly. For non-ideal systems, calculation of the equation of state is in principle possible, but exceedingly difficult in practice. Similar comments also apply to the calculation of key thermodynamic quantities such as the internal energy, U (Section 5.2) and the entropy, S (Section 10.4).

meaning that the statistical averages can be computed exactly...

We can nevertheless make *qualitative* predictions in the non-ideal case, as discussed below.

Note that the free particle assumption and the point particle assumption are independent—leading, e.g., to two separate vanishing terms in Equation (5.1). When both are in effect, this is described as the *kinetic model*, or alternatively, as an "ideal gas of point particles." All ideal gases—whether consisting of point particles or not—give rise to the same ideal gas equation of state [Equation (4.5)].

Hence, diatomic N_2 and-monoatomic Ar are equally well described by Eq. (4.5).

However, the ideal gas expressions for U (Section 5.4) and for S (Section 11.4) depend on whether or not the free particles are also point particles (Section 16.2). Point particles are presumed throughout this book, unless explicitly stated otherwise [apart from the obvious exception of Equation (4.5) itself].

The ideal gas assumption that $E_{P,i} = 0$ is rather crude; a more realistic case is portrayed in Figure 5.2—representing the intermolecular potential energy, $E_{P,i}$, experienced by molecule i, as a function of its distance, r, from another molecule. For simplicity, we assume point particles—so that particle orientation plays no role, and only r matters.

At very long ranges, where $r \to \infty$ and therefore the molar volume $V_m \to \infty$, the particles do not interact at all; they are effectively free particles, giving rise to ideal gas behavior. At intermediate ranges, attractive forces dominate; the attraction leads to P, Z, and U values that are

The intermolecular force is given by $F_i(r) = -dE_{P,i}(r)/dr$.

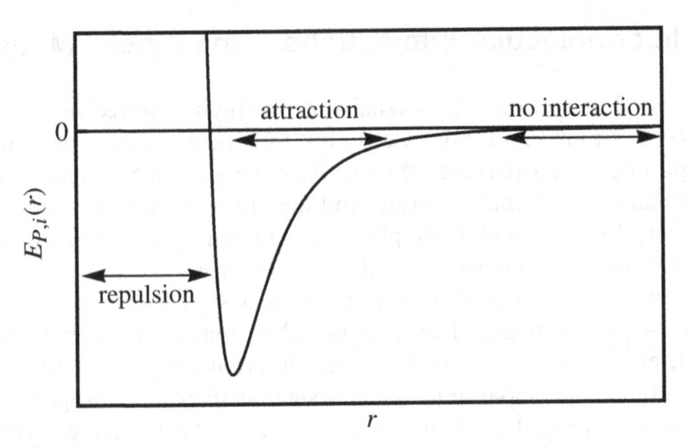

Figure 5.2 Intermolecular interaction potential. Typical intermolecular interaction potential energy, $E_{P,i}$, for a single point particle, i, as a function of distance, r, from another point particle. At close range, the particle feels the very steep "repulsive wall." At intermediate ranges, there is an intermolecular attraction. At very long ranges, the particles do not interact at all, and are effectively "free."

smaller than the ideal values. At very close range, $E_{P,i}$ is large and positive, and the two particles feel very strong repulsive forces pushing them apart. Macroscopically, this manifests as larger-than-ideal P, Z, and U values, as $V_m \to 0$.

The attractions and repulsions stem from *electrostatic interactions* among the electrons and protons.

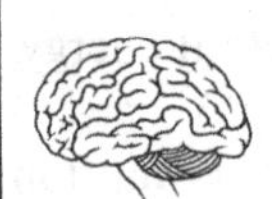

⊳⊳⊳ **To Ponder...** The near-impossibility of pushing two particles together closer than a certain minimum distance gives rise to the concept of particle *size*.

5.4 Equipartition Theorem & Temperature

According to the kinetic model, for an ideal gas of point particles, the only nonvanishing contribution to Equation (5.1) is the kinetic energy contribution,

Remember that the more important quantities such as 'm' are defined on p. xvii of the Textbook Guide.

$$E_{K,i} = \frac{m}{2}\left(v_{x,i}^2 + v_{y,i}^2 + v_{z,i}^2\right). \tag{5.4}$$

Substitution of Equation (5.4) into Equation (5.3) thus leads to:

$$\begin{aligned} U &= \sum_{i=1}^{N} \langle E_{K,i} \rangle \\ &= \sum_{i=1}^{N} \left[\frac{m}{2}\left\langle v_{x,i}^2\right\rangle + \frac{m}{2}\left\langle v_{y,i}^2\right\rangle + \frac{m}{2}\left\langle v_{z,i}^2\right\rangle\right] \qquad \text{[ideal gas]} \end{aligned} \tag{5.5}$$

There is an important result (end of Section 6.2) from statistical mechanics that states the following:

Equipartition Theorem: *For a system in thermal equilibrium, the statistically averaged kinetic energy contribution from each velocity coordinate is the same.*

In other words, every term in the sum in Equation (5.5) has the same value; every coordinate carries the same amount of kinetic energy, on average. This is true regardless of the velocity component, x, y, or z, or of the particular molecule, i—or even of the *type* of molecule. This is a profoundly important result.

Heavier particles thus move more slowly, on average (Sec. 6.3).

Since the above averages all have the same value, and since this value remains constant over time, we can regard it to be a true thermodynamic quantity. This is, in fact, what we use to define the *temperature*, as follows:

Definition 5.2 *For a system in thermal equilibrium, the* temperature, *denoted 'T', is defined in relation to the average kinetic energy of each velocity coordinate, as follows:*

$$\frac{kT}{2} = \frac{m}{2}\left\langle v_{x,i}^2 \right\rangle = \frac{m}{2}\left\langle v_{y,i}^2 \right\rangle = \frac{m}{2}\left\langle v_{z,i}^2 \right\rangle, \text{ for all } i. \quad \text{[thermal equilibrium]} \tag{5.6}$$

In other words, every molecular velocity coordinate carries the same amount of kinetic energy, on average, that is equal to $kT/2$.

▷▷▷ **Try It!!** How does *your* textbook explain temperature? Most of the non-physics reference texts do not provide a true *definition* of temperature, in the sense of Definition 1.1 (p. 4). The better treatments limit the definition to ideal gases only; the others merely provide a *description* of temperature, or list some of its qualities.

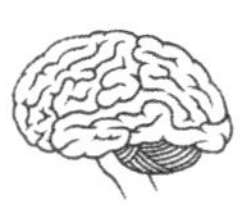

▷▷▷ **To Ponder...** Note from Equation (5.4) that kinetic energy is always *positive*. This is the reason why temperature, also, has an "absolute zero"; in the $T \to 0$ limit, the motion of all individual molecules becomes frozen.

From Equations (5.5) and (5.6), we obtain the $U(T, V)$ *state function* (see Section 8.1) for the ideal gas of point particles:

$$U(T, V) = \frac{3}{2}NkT = \frac{3}{2}nRT \qquad \text{[ideal gas]} \tag{5.7}$$

For more general ideal gases, there are also rotational coordinates that contribute to the molecular kinetic energy, and to which the equipartition theorem also applies. Thus, the general (rotating) ideal gas expression is

$$U(T, V) = \frac{d}{2} nRT, \qquad \text{[ideal gas (rotating)]} \quad (5.8)$$

where d is the total number of translational and rotational coordinates per molecule.

CORRESPONDENCE: Molecule Type & Number of Coordinates

Molecule type	**Translation**	**Rotation**	**Total (d)**
monoatomic (point, e.g. Ar)	3	0	3
diatomic (linear, e.g. N_2)	3	2	5
polyatomic (nonlinear, e.g. H_2O)	3	3	6

Note that for ideal gases, the state function $U(T, V)$ depends *only* on the thermodynamic variable T, not on V. This is due to the lack of intermolecular interactions—as a consequence of which, the intermolecular distance r, and thus the molar volume V_m, have no effect on the energy. Thus, for ideal gases, U and T are basically the same "thermodynamic energy" quantity—with U being the extensive version, and T the intensive version.

nor on P, if working with $U(T, P)$, though this choice of independent variables is less natural (Sec. 9.2)...

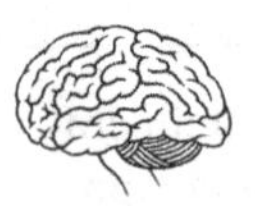

▷▷▷ **To Ponder...** Why then, do we need to define two separate quantities, U *and* T? This is because these are very different for *non*-ideal systems. Specifically, T relates only to molecular kinetic energy, whereas U also incorporates molecular potential energy.

Recall that $U(T, V)$ definitely *does* depend on V in the non-ideal case—in a manner that can be qualitatively understood, as already described in Section 5.3.

Chapter 6

Boltzmann Distribution & the Kinetic Model

Anderson: n/a; **Atkins:** Sec. F.5, p. 568, Sec, 20.1; **Atkins-life:** Sec. F.3(b), Sec. 1.2(c); **Baierlein:** p. 93, pp. 311–314, Sec. 13.3; **Callen:** Sec. 16-1, Sec. 16-10, Sec. 16-11; **Cengel:** n/a; **Chang:** Sec. 2.7, p. 52; **Elliott:** n/a; **Engel:** pp. 311–313, Sec. 30.2, Sec. 31.8, Sec. 33.2, Sec. 33.3; **Faure:** n/a; **Kittel:** pp. 58–64, pp. 391–395, pp. 419–421; **Levine:** Sec. 14.4, Sec. 14.5; **McQuarrie:** pp. 692–697, Sec. 25–2, Sec. 25–3; **Moran:** n/a; **Prausnitz:** pp. 758–760; **Reif:** Sec. 6 · 2, Sec. 6 · 3, Sec. 7 · 5, Sec. 7 · 9, Sec. 7 · 10; **Sandler:** n/a; **Schroeder:** p. 13, Sec. 6.1, Sec. 6.3, Sec. 6.4; **Silbey:** Sec. 16.1, Sec. 17.1, Sec. 17.2, Sec. 17.3; **Smith:** n/a; **Tinoco:** pp. 152–156, pp. 158–162

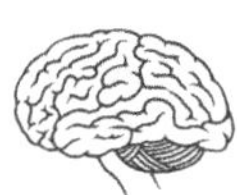

▷▷▷ **To Ponder...** Warning! In this chapter we delve into statistical mechanics. This material may be quite advanced, and/or not immediately relevant for some students. Though many may well find this chapter to be interesting and useful, it is not strictly necessary to master all of it now, in order to understand the rest of this book (although some of the results derived here are used in Chapters 5 and 11). Accordingly, some readers may choose to move directly on to Chapter 7.

6.1 Boltzmann Distribution

The kinetic model assumes that the point particles, of which the thermodynamic system is comprised, move around freely—i.e., without interacting—*except* when they collide with each other. These collisions are crucial, because they allow for a transfer of energy (and momentum) among the particles, which is necessary in order for the system to achieve equilibrium.

Throughout this chapter, an ideal gas of point particles is presumed.

A Conceptual Guide to Thermodynamics, First Edition. Bill Poirier.

Companion website: http://www.conceptualthermo.com

Even after the system has reached equilibrium, particle collisions continue to take place—giving rise to incessant changes in the individual particle states. Thus, over time, the position, (x_i, y_i, z_i), and velocity, $(v_{x,i}, v_{y,i}, v_{z,i})$, of a given particle i, do not remain constant, but take on a range of values. Through time averaging (see Section 5.2), a *probability distribution* can be defined, representing the relative amount of time that the particle spends in any given state.

Remember that the (x_i, y_i, z_i) and $(v_{x,i}, v_{y,i}, v_{z,i})$ values define the *single-molecule state* for particle i.

In most cases, all of the positions that lie within the system apparatus are accessible to any given particle, and equally probable (see Section 11.2). The situation is a bit more complicated for the velocity states, $(v_{x,i}, v_{y,i}, v_{z,i})$, however, because of energy conservation—whatever kinetic energy is gained by one particle in a single collision must be lost by its colliding partner. Particles with more kinetic energy are thus more likely to give up some of that energy in collisions. Accordingly, each particle spends less time in the higher-kinetic-energy states—as should be reflected in the equilibrium energy probability distribution.

From the Texts: A nice discussion may be found in **Engel**, pp. 311–313, and Sec. 30.2.

In fact, this distribution is very well known; it was derived by Boltzmann long ago, using statistical mechanics.

Definition 6.1 *For a system in thermal equilibrium, the relative amount of time spent in a molecular state with energy E is given by the* Boltzmann distribution, *denoted '$f(E)$', which takes the following form:*

From the Texts: Physics authors often refer to $f(E)$ as the *Boltzmann factor*.

$$f(E) = \exp(-E/kT) \qquad \text{[thermal equilibrium]} \qquad (6.1)$$

The Boltzmann distribution as presented in Equation (6.1) is not normalized—meaning that a sum of $f(E)$ over all molecular states does not equal one. Note that for a given T, there is indeed a decrease in probability with increasing E—a very sharp exponential decrease, in fact. Note also that for larger T, a greater effective range of E values—and therefore molecular states—is available. This aspect will be discussed further in Section 6.2.

6.2 Maxwell-Boltzmann Distribution

One of the nice features of Equation (6.1) is that it can be applied equally well to the molecular state of the whole system, or to the molecular state of a single particle ($E \to E_i$), or even to a single molecular *coordinate* [e.g., $E \to (m/2)v_{x,i}^2$]. In the second (single-particle) context, it is clear that each individual particle i is described by the same probability distribution; in the third context, we learn that even individual velocity *components*, such as $v_{x,i}$, are also all described by the same distribution.

provided that the system is *noninteracting*—i.e., ideal...

The third context above—i.e., of individual velocity components—is particularly important; it gives rise to the *Maxwell-Boltzmann distribution*:

$$\rho(v_x) = \left(\frac{m}{2\pi kT}\right)^{1/2} \exp(-mv_x^2/2kT) \qquad \text{[Maxwell-Boltzmann distribution]} \qquad (6.2)$$

In Equation (6.2) above, the 'x' component is explicitly considered, and the 'i' subscript has been dropped for convenience—though it is understood that the distribution applies equally well to each velocity component of every particle. The distribution has also been normalized, so that $\int \rho(v_x)\, dv_x = 1$.

For any molecular quantity that can be expressed in terms of v_x, the statistical average is the integral of that quantity times $\rho(v_x)$. For example, from Equation (6.2), we find that

$$\langle v_x \rangle = \int_{-\infty}^{\infty} v_x \rho(v_x)\, dv_x = 0, \qquad \text{[Maxwell-Boltzmann mean]} \quad (6.3)$$

implying that the mean particle velocity is zero, regardless of T. This does *not* mean that the individual particles are not moving—only that they are not moving in a statistically preferred direction.

See the To Ponder on p. 27.

The Maxwell-Boltzmann distribution also predicts that the most probable v_x states are those clustered around $v_x = 0$. Technically speaking, all v_x values in the full range, $-\infty < v_x < +\infty$, are theoretically possible. In practice, however, the likelihood of observing a given v_x value drops very sharply, as $|v_x|$ is increased beyond a certain point. As a consequence, the "effective" range of v_x is finite. There is also a marked temperature dependence, with larger T values leading to broader v_x ranges, as might be expected.

All of this can be seen clearly in Figure 6.1—a plot of the Maxwell-Boltzmann distribution, representing probability (or relative fraction of system particles) as a function of v_x, for two different temperatures.

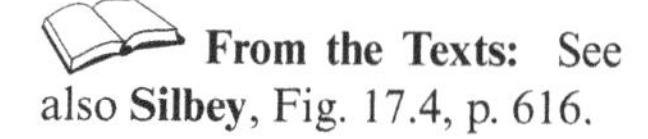

From the Texts: See also **Silbey**, Fig. 17.4, p. 616.

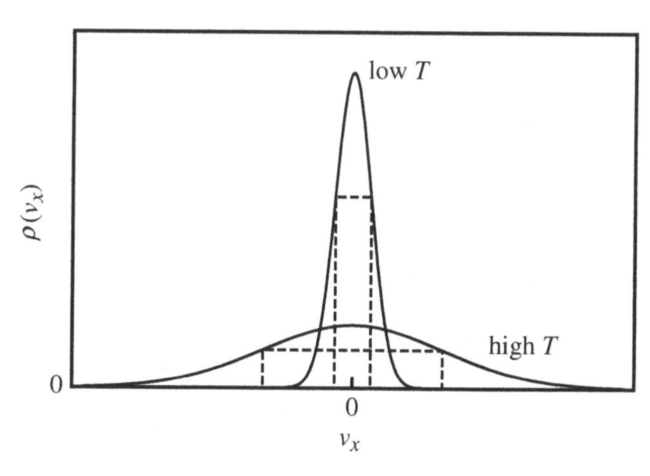

Figure 6.1 **Maxwell-Boltzmann distribution.** Maxwell-Boltzmann distribution for a single velocity component, v_x, for a single particle (the v_y and v_z distributions are the same, as are those for all other system particles with the same mass). The two solid curves correspond to different values of the temperature, T. As T increases, the distributions become broader, thereby encompassing a greater effective range of available velocity states (represented by the dashed line segments).

Statistical theory offers a very convenient quantitative measure of the effective range of a probability distribution, in the form of the *standard deviation*, denoted 'σ', and defined for v_x as follows:

From the Texts: Some authors use 'Δ', which should not be confused with the thermodynamic change 'Δ' of Sec. 7.2.

$$\sigma_{v_x} = \sqrt{\langle v_x^2 - \langle v_x \rangle^2 \rangle} \qquad \text{[standard deviation]} \quad (6.4)$$

Substitution of Equation (6.2) into Equation (6.4) yields σ_{v_x} for the Maxwell-Boltzmann distribution:

$$\sigma_{v_x} = \sqrt{\int_{-\infty}^{\infty} v_x^2 \, \rho(v_x) \, dv_x}$$

$$= \left(\frac{kT}{m}\right)^{1/2} \qquad \text{[Maxwell-Boltzmann standard deviation]} \quad (6.5)$$

Equation 6.5 is important for understanding the entropy of the ideal gas (see Section 11.3).

Finally, from the Maxwell-Boltzmann distribution, the equipartition theorem may be easily derived. By inspection of Equation (6.5), we immediately see that integrating $v_x^2 \, \rho(v_x)$ to obtain $\langle v_x^2 \rangle$ yields Equation (5.6).

Note that the mass factor in Eq. (6.5) is cancelled by that in Eq. (5.4).

6.3 Maxwell Distribution of Speeds

Our last distribution is the *Maxwell distribution of speeds*. This is obtained as follows. First, the three Maxwell-Boltzmann distributions for v_x, v_y, and v_z, are combined together to form a single, single-particle, three-dimensional distribution over the velocity vectors, (v_x, v_y, v_z). Next, the velocity vectors are expressed in spherical coordinates as (v, θ, ϕ), and the distribution is integrated over the two angular coordinates, (θ, ϕ). This results in a probability distribution solely in terms of the *speed*, v, or magnitude of the velocity vector,

Try It!! Try to derive Eq. (6.6) on your own, following the recipe described here.

$$v = \sqrt{v_x^2 + v_y^2 + v_z^2}.$$

The resultant probability distribution is:

$$\rho(v) = 4\pi v^2 \left(\frac{m}{2\pi kT}\right)^{3/2} \times \exp(-mv^2/2kT) \qquad \text{[Maxwell distribution of speeds]} \quad (6.6)$$

▷▷▷ ***Don't* Try It!!** Don't assume that the mean *speed* is zero for an ideal gas of point particles, just because the mean *velocity* is zero. In fact, since v must be nonnegative, $\langle v \rangle > 0$.

The Maxwell mean speed can be easily calculated:

$$\langle v \rangle = \int_0^\infty v\rho(v)\,dv = \left(\frac{8kT}{\pi m}\right)^{1/2} \qquad \text{[Maxwell mean speed]} \qquad (6.7)$$

However, the *root-mean-square (RMS) speed* is more commonly used, in practice:

$$v_{\rm rms} = \sqrt{\langle v^2 \rangle} = \sqrt{\int_0^\infty v^2\rho(v)\,dv} = \left(\frac{3kT}{m}\right)^{1/2} \qquad \text{[Maxwell RMS speed]} \qquad (6.8)$$

For N_2 (air) at $T = 300$ K, $v_{\rm rms} \approx 500$ m/s!

From Equations (6.7) and (6.8), we see that $v_{\rm rms}$ is slightly larger than $\langle v \rangle$. The most important conclusion, however, is that both speeds increase with temperature as $\sqrt{T}$. The effective range (standard deviation) of the Maxwell distribution also increases with temperature.

All of these trends can be clearly seen in Figure 6.2.

In addition to the temperature trends discussed above, the *mass trends* are also important. You will notice that in every equation in Sections 6.2 and 6.3 where T appears, it does so as the ratio (T/m). This means that the mass trends are exactly the opposite of the temperature trends. Thus, for a given temperature, heavy particles move more slowly than light particles (as discussed already in the marginal note on p. 39), and also have fewer velocity states available to them.

Try It!! Try verifying that this claim is indeed true, by inspecting all of the equations of the last two sections.

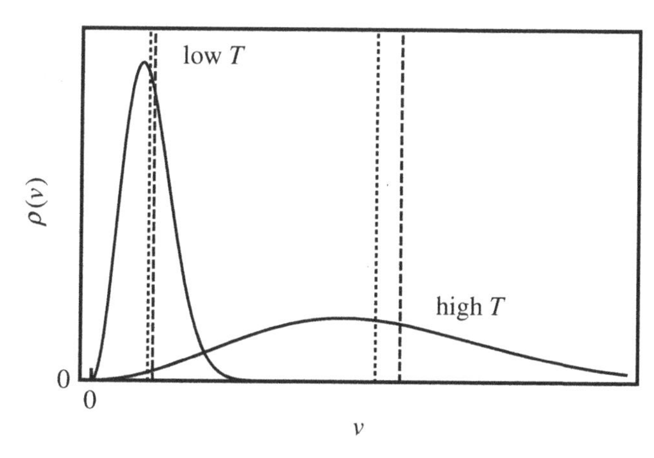

Figure 6.2 **Maxwell distribution of speeds.** Maxwell distribution of speeds v, for a single particle. This is obtained by integrating the three-dimensional Maxwell-Boltzmann distribution over the two angles. The two solid curves correspond to different values of the temperature, T. The mean speed, $\langle v \rangle$ (vertical dotted lines), is nonzero, and increases with T. The RMS speed, $v_{\rm rms} = \langle v^2 \rangle^{1/2}$ (vertical dashed lines), is larger than $\langle v \rangle$, but behaves similarly.

The Maxwell, Maxwell-Boltzmann, and Boltzmann distributions are all extremely important in their own right—the first two forming the basis of the kinetic theory of gases, and the last being the most important result in all of statistical mechanics! For our purposes, however, the most relevant result in this chapter is Equation (6.5)—which leads both to the equipartition theorem in Section 5.4, and to the ideal gas entropy in Section 11.3.

Part III

Thermodynamic Change

"Arguments based on heat engines have little appeal to chemists."
—M. L. McGlashan

"The majority of the phenomena studied in biology, meteorology, astrophysics, and other subjects are irreversible processes which take place outside the equilibrium state."
—Ilya Prigogine

"What politicians want to create is irreversible change because when you leave office someone changes it back again."
—Estelle Morris

A Conceptual Guide to Thermodynamics, First Edition. Bill Poirier.

Companion website: http://www.conceptualthermo.com

Chapter 7

First Law & Thermodynamic Change

Anderson: Sec. 2.6, Sec. 2.7, Sec. 3.4; **Atkins:** p. 44, Sec. 2.1, Sec. 2.2; **Atkins-life:** Sec. 1.1, pp. 23–24, Sec. 1.5; **Baierlein:** Sec. 1.3; **Callen:** Sec. 1-6; **Cengel:** Sec. 1–3, Sec. 2–6; **Chang:** p. 7, Sec. 3.2; **Elliott:** p. 39, Sec. 2.6, pp. 53–54; **Engel:** p. 6, Sec. 2.1; **Faure:** p. 155, Sec. 11.2; **Kittel:** p. 49; **Levine:** p. 37, Sec. 2.4; **McQuarrie:** p. 766, Sec. 19–3; **Moran:** Sec. 1.2, Sec. 2.5; **Prausnitz:** Sec. 2.1; **Reif:** Sec. 2 · 6, Sec. 2 · 7; **Sandler:** p. 4, Sec. 1.5, Sec. 3.1; **Schroeder:** p. 17, p. 19, pp. 123–124; **Silbey:** Sec. 1.1, Sec. 2.2; **Smith:** p. 12, Sec. 2.3; **Tinoco:** pp. 13–14

The First Law of Thermodynamics can be regarded as the macroscopic version of the conservation of energy law of physics. From the start, however, you are advised to keep the following Helpful Hint in mind.

⊳⊳⊳ **Helpful Hint:** Because thermodynamic systems are usually in mechanical and/or thermal contact with their surroundings, the First Law only manifests when there is a *change* in the thermodynamic state.

7.1 System & Surroundings

As discussed in Section 5.1, the molecular state energy E is conserved (constant over time) if the system is isolated. An *isolated system* is one that has neither mechanical nor thermal contact with its surroundings—it might as well be "alone in the universe."

Of course, this is not a very realistic picture. Generally speaking, contact between system and surroundings allows for changes not only of the

A Conceptual Guide to Thermodynamics, First Edition. Bill Poirier.

Companion website: http://www.conceptualthermo.com

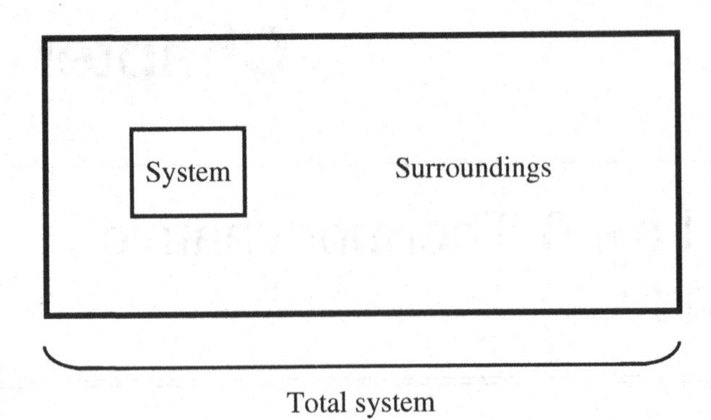

Figure 7.1 **System, surroundings, and total system.**

molecular state of the system (as in Section 5.1), but also its thermodynamic state—and that of the surroundings as well.

It can be convenient to consider the surroundings explicitly, as in Figure 7.1. Treated as its own system, the surroundings has its own thermodynamic variables and quantities, denoted with a 'sur' subscript. Together, the system plus surroundings form the *total system*, whose quantities adopt a 'tot' subscript.

Thus, 'X' for the system, 'X_{sur}' for the surroundings, 'X_{tot}' for the total system.

The picture above resembles the subsystems picture of Section 4.3, but with some key differences. Whereas subsystems are usually treated "equally," the system and its surroundings are not. For instance, the surroundings are usually much larger than the system, as in the case of a heat bath (see Section 9.1).

Another difference is the requirement that the *total system must be isolated.* Arguments that apply only to isolated systems (such as energy conservation) may then be referred to the total system—allowing the system itself to be more general. Of course, the precise division into "system" and "surroundings" is flexible.

From the Texts: Some physics authors even take the total system to be the *entire universe*! At least this is isolated...

▷▷▷ **Helpful Hint:** As with independent variables, you are free to choose the "system" and "surroundings" for a given thermodynamics problem however you wish. Choose wisely—i.e., so as to make the problem easier to solve—and remain consistent in your choice from start to finish.

7.2 Thermodynamic Change

If "nothing ever happened" (see Talking Heads excerpt on p. 9), life would be pretty boring. This is the case with thermodynamic systems in equilibrium, so long as there is no change in external factors (Definition 4.1, p. 26). What happens when external factors *do* change? In most cases, this gives rise to a *change of thermodynamic state*. One obvious consequence of such a change is that at least one of the two independent variables must change its value.

We will often refer to the initial thermodynamic state as "state A," and denote the initial values of thermodynamic quantities with an 'i' subscript. Likewise, "state B," and 'f' subscripts, are used for the final state. The change in the value of a generic quantity X, under the thermodynamic change, is then given as

$$\Delta X = (X_f - X_i). \tag{7.1}$$

Note that *most* thermodynamic quantities—not just the variables—may be expected to change their values under a thermodynamic change. The surroundings quantities also undergo their own change.

EXAMPLE: Gas Expansions

Gas expansions are a standard topic in any thermodynamics course. This type of thermodynamic change requires that a wall be either moved or removed, resulting in an expansion (or compression) of the system size. Despite sounding very simple, gas expansions can in fact be surprisingly complex.

In any case, all gas expansions share one obvious fact in common: a change in the system volume variable, V.

- initial volume, $V = V_i$.
- final volume, $V = V_f$.
- *change* in volume, $\Delta V = (V_f - V_i)$.
- true "expansion" requires $V_f > V_i$.

▷▷▷ **Helpful Hint:** The opposite of a true $\Delta V > 0$ gas expansion—i.e., the case where $\Delta V < 0$—is a gas *compression*. Many authors loosely extend the term "gas expansion" to include the compression case as well.

Science Doesn't Care; see Sec. 8.5 and Chap. 16.

From the Texts

One of the major reasons why thermodynamic processes can be so complicated is that there are *two* independent variables, rather than only one. Consequently, it is never sufficient to rely on a single variable only. An unambiguous description of a thermodynamic change requires a specification of what *both* independent variables are doing. That said, thermodynamic changes often occur under the special condition that one of the variables is held fixed (constant).

▷▷▷ **Helpful Hint:** *Always* choose the constant or fixed variable as one of your two independent variables; this is invariably the easiest way to solve a given thermodynamics problem.

EXAMPLE: Gas Expansion at Constant Pressure

By stipulating that a gas expansion occurs *at constant pressure*, we provide the information necessary to specify the thermodynamic change exactly—i.e., to uniquely determine the initial and final states.

The natural choice for the two independent variables includes the fixed variable P, and the changing variable V. Initial and final thermodynamic states are determined from initial and final values for both independent variables:

Variable	Initial value	Final value	Change
P	$P_i = P$	$P_f = P$	$\Delta P = 0$
V	V_i	V_f	$\Delta V = (V_f - V_i)$

▷▷▷ **Helpful Hint:** Since the pressure is held fixed in this example, it is not actually necessary to refer to 'P_i' and 'P_f'; we can simply use 'P' for both.

Thermodynamic changes can be represented using *indicator diagrams*—two-dimensional plots, in which each axis represents a different independent variable, and each point a different thermodynamic state. Indicator diagrams are often used to show the path taken through "state space," to get from A to B, as in Figure 7.2.

7.3 First Law

We now have all the tools needed to derive the First Law of Thermodynamics. Because the total system is isolated, E_{tot} is conserved. The same must also be true of $U_{\text{tot}} = \langle E_{\text{tot}} \rangle$ (ignoring the internal state energy; see Section 5.2). Also, U is an extensive quantity. This means that the total quantity value is equal to the sum of the values for each of its parts—i.e., $U_{\text{tot}} = U + U_{\text{sur}}$. This leads to the

First Law (total system): *The internal energy of the total system is conserved under any thermodynamic change:*

$$\Delta U_{\text{tot}} = \Delta U + \Delta U_{\text{sur}} = 0 \qquad \text{[always]} \qquad (7.2)$$

Whatever energy is gained by the system must be lost by the surroundings, and vice-versa.

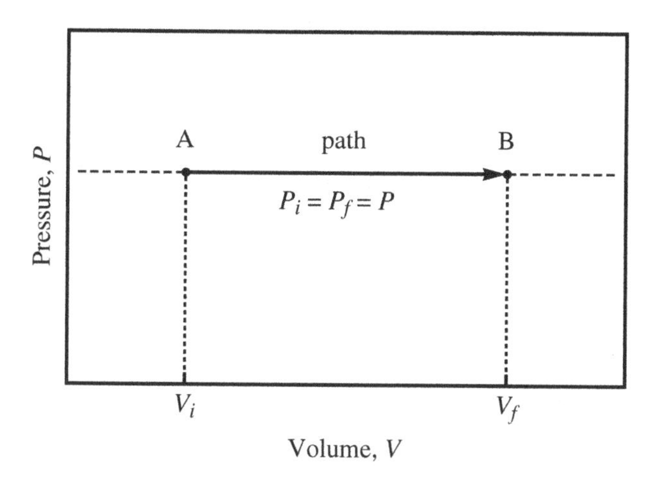

Figure 7.2 Gas expansion at constant *P*. Indicator diagram for the gas expansion at constant pressure example discussed in the box on p. 52. The system begins in the initial state A at ($P_i = P$, V_i). It then undergoes a thermodynamic change to the final state B at ($P_f = P$, V_f), following the constant pressure path indicated.

In its most fundamental form above, the First Law is a statement about the *total system*. In order to convert Equation (7.2) into its more familiar, system-based form, we must consider the ways in which thermodynamic energy can be transferred between system and surroundings. As introduced in Section 4.3 (and to be further elaborated on in Section 8.2): mechanical contact leads to energy transfer in the form of *work*, denoted 'W'; thermal contact leads to energy transfer in the form of *heat*, denoted 'Q'. For our purposes, these are the only means of energy transfer between system and surroundings. This leads to the revised

There are also other kinds of energy transfer that can occur, e.g., electrical and chemical work, associated with the other forms of "contact" discussed on pp. 28–29.

First Law: *Under any thermodynamic change,*

$$\Delta U = Q + W. \qquad \text{[always]} \quad (7.3)$$

Note that *positive* Q and W lead to *increased* U, reflecting a focus on the system. Engineers often focus on the *surroundings*, leading to the opposite *sign convention* (at least for W). Note also that Equations (7.2) and (7.3) imply the following useful relations:

From the Texts: See also the discussion on p. xviii. Like IUPAC, some engineers are coming around (see, e.g., **Elliott**, **Sandler**, and **Smith**).

$$Q_{\text{sur}} = -Q \qquad (7.4)$$

$$W_{\text{sur}} = -W \qquad (7.5)$$

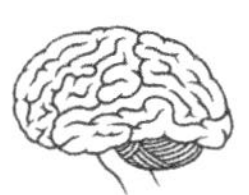

⊳⊳⊳ **To Ponder...** Thermodynamic conservation of energy—i.e., the First Law—depends crucially on the idea of thermodynamic *change*. For a system in equilibrium, it is true that U is conserved—but so is *every other* state function! On the other hand, *only* for U is Equation (7.2) always true—i.e., in general, $\Delta X_{\text{tot}} = \Delta X + \Delta X_{\text{sur}} \neq 0$.

Chapter 8

Work, Heat, & Reversible Change

Anderson: Sec. 2.6.1, Sec. 2.6.2, Sec. 3.3, Sec. 3.4, Sec. 3.6; **Atkins:** Sec. 2.1, Sec. 2.3; **Atkins-life:** Sec. 1.2, Sec. 1.3, p. 69; **Baierlein:** Sec. 1.3, Sec. 1.6, p. 56, Sec. 3.6; **Callen:** Sec. 1-8, Sec. 4-1, Sec. 4-2, Sec. 4-5; **Cengel:** Sec. 2–3, Sec. 2–4, Sec. 2–5, Sec. 6–6; **Chang:** Sec. 3.1; **Elliott:** p. 17, Sec. 2.1, Sec. 2.4, Sec. 2.6; **Engel:** Sec. 2.2, Sec. 2.3; **Faure:** Sec. 11.2; **Kittel:** p. 64, pp. 240–244; **Levine:** pp. 29–30, p. 41, Sec. 2.2, Sec. 2.3, Sec. 2.8, Sec. 2.10; **McQuarrie:** Sec. 19–1, Sec. 19–2, Sec. 19–6; **Moran:** Sec. 2.1, Sec. 2.2, Sec. 2.4, Sec. 5.3; **Prausnitz:** Sec. 2.1; **Reif:** Sec. 2 · 9, Sec. 2 · 10, p. 82, Sec. 4 · 1, Sec. 4 · 2; **Sandler:** Sec. 1.5, Sec. 3.1, Sec. 4.2; **Schroeder:** Sec. 1.4, Sec. 1.5, pp. 82–83; **Silbey:** Sec. 2.1, Sec. 2.4; **Smith:** Sec. 1.7, Sec. 1.9, Sec. 2.8; **Tinoco:** pp. 15–21, pp. 26–28, p. 35, pp. 157–158

8.1 State Functions & Path Functions

Work, W, and heat, Q, are obviously very important in thermodynamics, playing a central role in the First Law [Equation (7.3)]. They are also unique, in being the only thermodynamic quantities that are not also *state functions*. By the latter, we mean quantities such as $U(T, V)$ that depend only on the thermodynamic state. In contrast, values for work and heat can only be assigned to thermodynamic *processes* (changes of state)—never to individual states themselves.

the only quantities that we will consider, anyway…

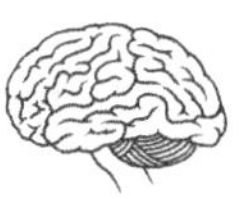

⊳⊳⊳ **To Ponder…** One imagines the stereotypical Hollywood agent, promoting his or her clients as "having heat." Thermodynamically speaking, this is nonsense; one can "release heat" or "absorb heat," but not "have heat." What one "has"—though admittedly a lot less catchy—is internal energy.

A Conceptual Guide to Thermodynamics, First Edition. Bill Poirier.

Companion website: http://www.conceptualthermo.com

For a change from initial state A to final state B, the generic state function change, ΔX, depends only on A and B themselves [as per Equation (7.1)], and not on the specific path taken through "state space" to get from A to B (as in Figure 7.2). The opposite is true for the non-state-function quantities, W and Q. In other words, it is not only the path *endpoints*, A and B, that determine W and Q, but also all of the states in between. Such quantities are called *path functions*, to emphasize this path dependence.

▷▷▷**Helpful Hint:** To use a driving analogy, "elevation" is a state function; it depends only on where you are, not on where you've been. On the other hand, "mileage" is a path function—nonzero for a round trip, and otherwise dependent on your past travel history.

▷▷▷ **Helpful Hint:** For equations describing thermodynamic change, *every term* must have one 'Δ' in it. The *only* exceptions are for the path functions, W and Q, as in Equation (7.3). Also, remember that 'Δ' expressions are always computed as "final minus initial" [Equation (7.1)]—not the other way around.

▷▷▷ ***Don't* Try It!!** Don't *ever* combine 'ΔX' expressions and 'X' expressions (i.e., with no 'Δ') in the same equation! This is the *third* most common error made by students, after the "liter/cubic meter" error of Section 3.2, and the "ideal gas" error of Section 4.4. Specific examples with correct usage are listed in the Helpful Hint on p. 74.

EXAMPLE: Cycles

Path functions can exhibit "surprising" behavior. For example, consider a loop or *cycle*—i.e., a path that begins and ends at the same state, $(P_i, V_i) = (P_f, V_f)$. For all state functions, $X(P, V)$, one must have $\Delta X = 0$ around the loop. W and Q, on the other hand, are generally *nonzero* around a cycle [although $W = -Q$, to satisfy Equation (7.3)]. Thus, one can generate as much work or heat as desired, simply by going around the loop an arbitrary number of times. Such behavior may seem "weird" at first, but is actually very important, and must be properly understood. Cycles will be discussed further in Sections 12.1 and 13.2 (see also Figure 13.2).

Cycles are very important in engineering applications.

without which, we would not have heat engines or air conditioners (see Chaps. 12 and 13)...

Science Doesn't Care; see Secs. 9.4 and 12.1.

8.2 Definition of Work

We all have some intuitive notions of what "work" and "heat" are. More mathematically, we learn from Equation (7.3) that both W and Q are extensive quantities with units of energy. But these are mere *descriptions*, not quantitative, scientific definitions. Fortunately, physics already provides a definition of mechanical work—one that is best expressed in *differential form*.

Consider an *infinitesimal* thermodynamic change—i.e., one in which the initial state A and final state B become arbitrarily close. Thus, $(P_f, V_f) \approx (P_i, V_i)$, or $(P_f, V_f) = (P_i + dP, V_i + dV)$, where dP and dV are *differentials*. For an infinitesimal change, $\Delta P \to dP$ and $\Delta V \to dV$; the same is also true for all other state function quantities—i.e., $\Delta X \to dX$.

except near a phase transition to a condensed phase (see Chap. 17)...

▷▷▷ **Helpful Hint:** For infinitesimal changes, '(P, V)' is often used for (P_i, V_i), and '$(P + dP, V + dP)$' for (P_f, V_f). This notation may obscure the fact that even an infinitesimal thermodynamic change is still a *change of state*, rather than just a single state. Don't be fooled!

***Don't* Try It!!**

For an infinitesimal change, the differential form version of Equation (7.3) is satisfied.

First Law (differential form): *Under any* infinitesimal *thermodynamic change,*

$$dU = dQ + dW. \qquad \text{[infinitesimal]} \quad (8.1)$$

According to physics, the infinitesimal work done when moving a macroscopic object against an opposing force is given by

$$dW = -F_{\text{sur}}\, dz, \qquad (8.2)$$

where F_{sur} is the opposing force, and dz is the infinitesimal change in the object's position.

One could just as easily use dx or dy.

In thermodynamic terms, the "opposing" force is that imparted by the surroundings, and the macroscopic object is the (movable) dividing wall. These ideas are conveniently encapsulated in the *piston-cylinder apparatus* of Figure 8.1.

From Equation (4.4), pressure can be obtained from force, by dividing by the area of the movable wall. Likewise, infinitesimal distance becomes infinitesimal *volume* when multiplied by the wall area, so that

$$dW = -P_{\text{sur}}\, dV. \qquad \text{[infinitesimal]} \quad (8.3)$$

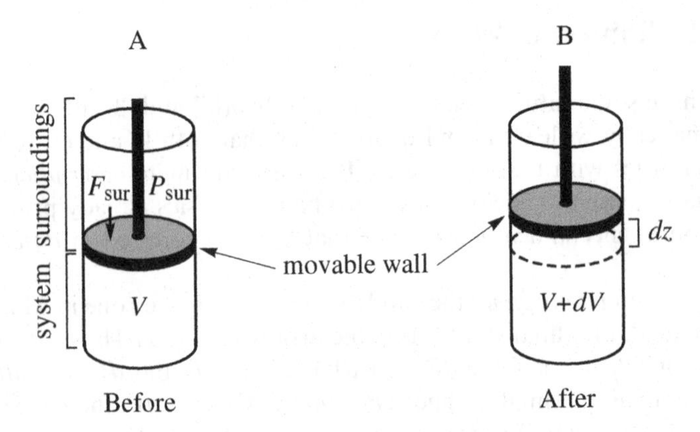

Figure 8.1 Expansion work for piston-cylinder apparatus. The "system" is the gas inside the cylinder. The piston itself serves as the movable dividing wall; as part of the surroundings, it also provides the opposing force, F_{sur} (and the surroundings pressure, P_{sur}). Expansion is in the vertical direction. The resultant infinitesimal work [Eqs. (8.2) and (8.3)] is negative; work is done *by* the system *on* the surroundings.

Integrating both sides of Equation (8.3) then results in

Definition 8.1 *For any thermodynamic process, the (expansion) work, denoted 'W', is defined to be*

$$W = -\int_{V_i}^{V_f} P_{sur}(V)\,dV, \qquad \text{[always]} \quad (8.4)$$

where $P_{sur}(V)$ specifies the path from A to B.

From Equation (8.4), there are two conditions that are absolutely required of any thermodynamic process, in order for it to exhibit nonzero work. These are:

1. a change in the system volume, V.
2. a nonzero surroundings pressure, P_{sur}, against which the expansion takes place.

Both conditions are necessary. Without Condition 1, there is no macroscopic motion, and therefore no work. Without Condition 2, there is *free expansion* (see Section 16.2), but again, no work.

▷▷▷**Helpful Hint:** Equation 8.4 is a *definition*, and as such may always be relied upon. You will encounter many other, conditional expressions for W (Sections 8.5 and 16.3) that you may also use—but which only apply under special circumstances. When in doubt, use the definition; it is *always* correct.

▷▷▷**Helpful Hint:** Unlike in Figure 7.2, in general it is the *surroundings* pressure, P_{sur}, rather than the system pressure, P, that defines the thermodynamic path (as a function of system volume, V). The reasons for this will be discussed in Section 8.4.

▷▷▷***Don't* Try It !!** Don't forget the minus sign in Equation (8.4)! The minus sign is needed to ensure that work is *negative* for a true $\Delta V > 0$ expansion. This is the *fourth* most common error made by students, after those discussed in Sections 3.2, 4.4, and 8.1.

▷▷▷***Don't* Try It !!** Don't assume that $P_{sur}(V)$ is a constant that can be pulled out of the integral in Equation (8.4); in general it is a function of V, that must be properly integrated. This mistake—or perhaps using P when P_{sur} should be used—is the *fifth* most common error made by students.

8.3 Definition of Heat

Heat will be treated a little differently than the other thermodynamic quantities considered thus far. Rather than *describe* heat at the macroscopic scale and *define* it at the molecular scale, we will do exactly the opposite.

Qualitatively, we already understand what heat is at the molecular scale: it is the transfer of molecular kinetic energy across a diathermic wall. This is merely a description, however, not a definition. Whereas it might be possible to actually define Q at the molecular scale, we prefer to define it at the *macroscopic* scale, directly from the First Law:

$$Q = \Delta U - W \qquad \text{[heat]} \qquad (8.5)$$

Thus, heat is "whatever is left over" in a thermodynamic change, when the work is subtracted from the change in the internal energy.

Since ΔU and W are well defined for any thermodynamic process, so is Q. There is absolutely nothing wrong with this definition—it satisfies the criterion of Definition 1.1 (p. 4) perfectly well. Moreover, from a practical standpoint, Equation (8.5) conveys the usual procedure that you should follow, when asked to compute Q in problems.

It's OK to be Lazy, in this case, provided that there are no other forms of energy transfer available…

8.4 Reversible & Irreversible Change

Why is the thermodynamic path taken to be $P_{sur}(V)$ rather than $P(V)$? This is because P itself is not necessarily well defined throughout the entire process, since the system may venture far from equilibrium at intermediate times. We know that at least the path endpoints, A and B, are necessarily both equilibrium states—but in between, it could well be another story.

From the Texts: Nevertheless, some authors try to define a $P(V)$-based "reversible work," even for *irreversible* processes. We frown on this; see the Helpful Hint on p. 94.

On the other hand, we do presume that the *surroundings* are always in equilibrium throughout the thermodynamic change—so that P_{sur} is always well defined. An extremely important consequence is that the work [Equation (8.4)] is also always well defined, even for irreversible processes (see below).

V is also always well defined, even for a system that is not in equilibrium.

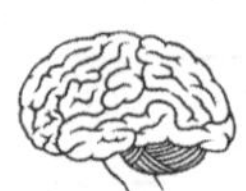

▷▷▷ **To Ponder...** The above underscores another important difference between system and surroundings—and another reason why the "total system" (Section 7.1) and "subsystems" (Section 4.3) pictures are not the same.

We distinguish two types of thermodynamic change:

reversible The system is in equilibrium at every step along the path from A to B, with $P = P_{sur}$ throughout.

irreversible The system is far from equilibrium for at least part of the path from A to B.

Reversible change is an idealization—if a system were *truly* in equilibrium (Definition 4.1, p. 26), its thermodynamic state would never change! In practice, what "reversible" really means is that external factors change *very slowly*—so that the system can gently readjust itself, incrementally throughout the process. As a consequence, the system never gets very far away from the equation of state.

From the Texts: Hence, some authors refer to "quasi-static change," rather than reversible change.

In contrast, irreversible change occurs after a *sudden* external change has pushed the system far from equilibrium. The system then undergoes an automatic or "spontaneous" thermodynamic change, until a new equilibrium state is reached (see Section 12.3).

COMPARISON: Reversible vs. Irreversible Change

Reversible	Irreversible
equilibrium throughout	equilibrium at endpoints only
path on equation of state	path far from equation of state
external change slow	external change fast
not spontaneous	spontaneous
maximum work	less than maximum work

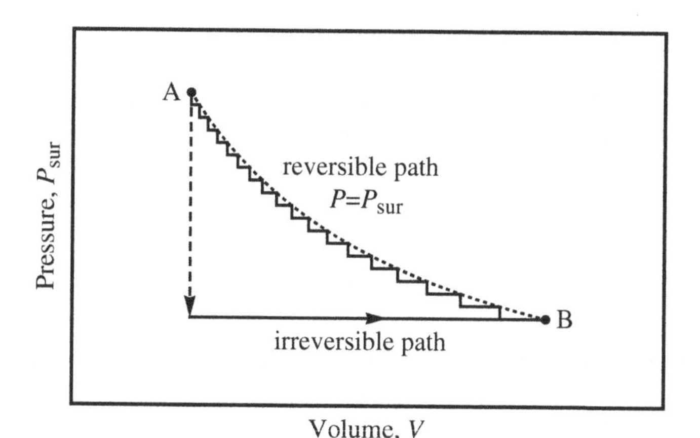

Figure 8.2 Reversible and irreversible paths. Two isothermal paths—one reversible and one irreversible—to get from A to B. The dotted curve represents the true reversible path, an isotherm. The vertical dashed line and horizontal solid line represent the irreversible path. See main text for further discussion.

Figure 8.2 depicts two different isothermal (constant T) paths to get from A to B—one a reversible path, the other an irreversible path. The reversible path is an *isotherm*—i.e., a contour of the $T(P, V)$ equation of state (treating T as the dependent variable; see box on p. 24). In reality, a nearly reversible path might look more like the staggered solid curve—obtained when P_{sur} is reduced *slowly*, so that $P \approx P_{sur}$ throughout. In contrast, when P_{sur} is *suddenly* reduced (vertical dashed line), a spontaneous change ensues (horizontal solid line) until equilibrium is again restored at B. This is the irreversible path.

Your primary textbook most likely discusses work as being "the area under the path." The two paths indicated in Figure 8.2 clearly have different W values. Note that the reversible path generates more work (larger $|W|$) than does the irreversible path. This is always the case for a true $\Delta V > 0$ expansion; of all physically realizable paths from A to B, the reversible path yields *maximum work*.

although it may not distinguish P from P_{sur}, as we have done here...

with some caveats, to be discussed in Sec. 14.3... This is why reversibility is so important to engineers.

Why should this be the case? To expand a gas, one must *first* ease up on P_{sur}, and *then* allow the system to increase V on its own (thereby restoring equilibrium). Hence, all real isothermal expansion paths lie *below* the reversible path indicated in Figure 8.2. To *compress* a gas, one first *increases* P_{sur}, and then allows V to *decrease*. Thus, all real compression paths lie *above* the reversible path in Figure 8.2.

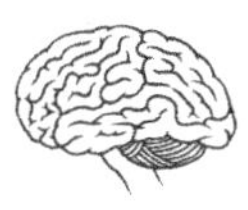

▷▷▷ **To Ponder...** Thus do we arrive at the real meaning of *reversible* and *irreversible*. Unbeknown to E. Morris (see quote on p. 47), "irreversible" does *not* mean that the change cannot be undone! No matter how the system gets from A to B, there is always *some* path that gets it back to A again. But the forward and reverse paths must be different—*except* in the case of the reversible path.

8.5 A Gas Expansion Example

There is a bewildering variety of different kinds of gas expansions. There are *reversible* and *irreversible* expansions, for gases that are either *ideal* or *non-ideal*. Some expansions are *isothermal* (constant T); others are *isobaric* (constant P_{sur}). There are also *free* ($W = 0$) and *adiabatic* ($Q = 0$) gas expansions.

In Section 16.3, we offer a comprehensive compendium. Here, we successively narrow the focus down to one classic example: the reversible isothermal expansion of an ideal gas. Refer to Figures 8.1 and 8.2.

Reversible expansion: For the special case of a *reversible* gas expansion, $P_{sur} = P$ at every point along the path from A to B. This means that we can replace the vertical axis in Figure 8.2 with P. The path $P_{sur}(V)$ thus becomes the *reversible* path, $P(V)$. Moreover, because the system is in equilibrium throughout, we know that the *reversible path lies on the equation of state*. Consequently, the temperature is not arbitrary, but is determined from the $T(P, V)$ equation of state $[T(V) = T(P(V), V)]$.

We actually did this once before, in Fig. 7.2—now seen to be a *reversible isobaric* expansion.

▷▷▷**Helpful Hint:** This situation is quite different from the isothermal case of Figure 8.2, for which all paths other than the dotted curve are *irreversible*. (See also Figure 13.1.)

For a reversible expansion process, the work becomes:

$$W = -\int_{V_i}^{V_f} P(V)\,dV \qquad \text{[reversible]} \qquad (8.6)$$

Reversible expansion of ideal gas: Let us further specialize, by assuming that our reversible expansion applies to an *ideal gas*. This means that the reversible path $P(V)$ in Equation (8.6) can be related to T via the ideal gas equation of state, $P(V) = nRT/V$. The work, in turn, becomes $W = -\int nRT(1/V)\,dV$.

▷▷▷***Don't* Try It!!** Don't assume that T is constant, and can therefore be pulled out of the integral above! In general, $T = T(V)$ is a function of V—related to the reversible path, $P(V)$, which is not yet specified. We can make no further progress until we actually specify this path.

Reversible isothermal expansion of ideal gas: *Only* by specifying that the reversible ideal gas expansion path is *isothermal* can we treat T as a constant. We thus obtain $W = -nRT\int(1/V)\,dV$, or

Remember: *two* independent variables must always be specified (here, the constant T and the varying V).

$$W = -nRT\,\ln\left(\frac{V_f}{V_i}\right). \qquad \text{[ideal gas, reversible, const } T\text{]} \qquad (8.7)$$

LOG BLOG: Helpful Hints about Logarithms

▷▷▷ **Helpful Hint:** When computing the work for a reversible isothermal ideal gas expansion, make sure to use the 'ln' key on your calculator (natural log), rather than the 'log' key (which usually means log base 10).

▷▷▷ **Helpful Hint:** Although you can multiply and divide dimensional units to get new "derived" units, you cannot take the *logarithm* of a unit. Therefore, the argument of a logarithm should always be *dimensionless* (although some authors do not always follow this convention...).

From the Texts

▷▷▷ **Helpful Hint:** You can take advantage of the previous Helpful Hint when solving problems. Specifically, there is often no need to convert units inside a logarithm, since the final units must cancel. The most common example is a ratio of two quantities with the same dimensions, such as the volume ratio in Equation (8.7). If, in this case, the problem gives you V_i and V_f in say, liters (L), there is no need to convert these to SI units (m^3) in order to obtain the final result for W in SI units (J).

Chapter 9

Partial Derivative Quantities

Anderson: Sec. 3.5; **Atkins:** Sec. 2.4, Sec. 2.5, Sec. 2.10, Sec. 2.11, pp. 91–93; **Atkins-life:** Sec. 1.4, Sec. 1.5, Sec. 1.6; **Baierlein:** Sec. 1.4, p. 242, Sec. 12.2; **Callen:** Sec. 3-1, Sec. 3-9, Sec. 5-2, Sec. 5-3; **Cengel:** Sec. 3–5, Sec. 4–3, Sec. 4–4, pp. 661–666, pp. 674–685; **Chang:** pp. 46–47, Sec. 3.3, Sec. 3.5; **Elliott:** pp. 19–20, pp. 52–53, Sec. 2.10, Sec. 6.2; **Engel:** Sec. 2.5, Sec. 2.6, Sec, 2.9, Sec. 3.1, Sec. 3.2, Sec. 3.3, Sec. 3.5; **Faure:** Sec. 11.3, Sec. 11.5; **Kittel:** pp. 62–63, p. 246, pp. 334–338; **Levine:** Sec. 1.6, Sec. 1.8, Sec. 2.5, Sec. 2.6; **McQuarrie:** pp. 683–690, Sec. 19–7, Sec. 19–8; **Moran:** Sec. 3.6, Sec. 3.9, Sec. 3.13, Sec. 11.5; **Prausnitz:** Sec. 2.1; **Reif:** Sec. 2 · 11, Sec. 3 · 6, Sec. 4 · 4, Sec. 5 · 2, p. 162, Sec. 5 · 7, Sec. 5 · 8; **Sandler:** p. 53, Sec. 3.3, pp. 187–207; **Schroeder:** Sec. 1.6, Sec. 3.2, p. 149; **Silbey:** Sec. 2.3, Sec. 2.6, Sec. 2.8, Sec. 2.13; **Smith:** Sec. 2.10, Sec. 2.11, p. 68, p. 201; **Tinoco:** pp. 24–26, pp. 30–31, p. 36, pp. 87–90, pp. 618–621

In the previous chapters (and in Chapter 16), we have explored essentially all there is to know about T, P, V, U, W, and Q, on their own. Accordingly, in this chapter, we introduce a host of new thermodynamic quantities. Instead of delving down to the molecular scale and applying statistical mechanics—as was done previously, e.g., for U—the new quantities are derived *directly* at the macroscopic scale, from previously defined quantities. This is perfectly legitimate, since **It's OK to be Lazy**; in fact, we have already used this strategy in our definition of heat [Equation (8.5)].

Most of the new quantities are defined as mathematical derivatives of other state function quantities—exploiting the fact that the latter are, in fact, functions of the independent thermodynamic variables. Of course, there are *two* variables, rather than just one—necessitating the use of *partial derivatives*, rather than ordinary derivatives, and otherwise making things more difficult and confusing.

ah well, **Science Doesn't Care**...

A Conceptual Guide to Thermodynamics, First Edition. Bill Poirier.

Companion website: http://www.conceptualthermo.com

9.1 Internal Energy & Heat Capacity at Constant Volume

In Section 8.5, we discussed the reversible isothermal expansion of an ideal gas. How could such an experiment be realized in practice? One simple scenario would be to imagine dunking the entire piston-cylinder apparatus of Figure 8.1 into a very large *heat bath*, represented schematically as the surroundings in Figure 7.1.

The term "bath" suggests liquid surroundings; liquids generally have much higher *heat capacities* than gases.

While the piston is slowly moved, and the system thus reversibly expanded, heat flows from the bath to the system. This compensates for the system internal energy that is being lost in the form of work. However, because energy is *extensive* and temperature *intensive*, the heat loss of the surroundings results in only a negligible reduction in T_{sur}—which is therefore effectively constant. Finally, since the system and surroundings are in thermal equilibrium with each other throughout the expansion process, $T = T_{sur}$ is also constant.

Remember that the heat bath is presumed to be huge, compared to the system.

The above scenario plays out the way it does because of something called the *heat capacity*—essentially, the ratio between heat and temperature change. You have no doubt known about heat capacity for some time, but probably not as a true thermodynamic state function. Our goal in this section is to define one such heat capacity quantity. First though, we must consider a new type of thermodynamic change.

Instead of changing V while keeping T constant (i.e., as in isothermal expansion), consider changing T while keeping V constant. Since the volume does not change, there is no work; all internal energy change is thus due to heat alone:

This is *not* an expansion process.

$$\Delta U = Q \qquad \text{[const } V\text{]} \quad (9.1)$$

From the Texts: Some authors prefer the terms "isochoric," "isovolumetric," or "isometric."

Constant V calorimetry is called *bomb* calorimetry.

We can therefore measure Q experimentally simply by measuring ΔU, or vice-versa—but only under *constant volume* conditions. For an ideal gas, moreover, we can obtain both Q and ΔU simply by measuring ΔT and the amount of substance. This is the basis of *calorimetry*:

$$\Delta U = Q = \frac{3}{2} nR\Delta T \qquad \text{[ideal gas, const } V\text{]} \quad (9.2)$$

▷▷▷**Helpful Hint:** When doing calorimetry problems, it is the *changes* in quantities that matter—i.e., ΔT and ΔU, *not* T and U themselves. Equation (9.2) is used for problems; however, it is the *differential form* (Section 9.4) that is used in the definition of the heat capacity.

Definition 9.1 *The* heat capacity at constant volume, *denoted 'C_V', is defined to be the partial derivative of $U(T, V)$ with respect to T at constant V—i.e.,*

$$C_V = \left(\frac{\partial U}{\partial T}\right)\bigg|_V \qquad \text{[always]} \quad (9.3)$$

Implicit in the partial derivative expression of Equation (9.3) is the assumption that T and V are the two independent variables—with T varying, and V fixed. Note that $C_V = C_V(T, V)$ is itself a state function quantity. In the general case, it depends on both variables, T and V. For the ideal gas special case, however, we see from the differential form of Equation (9.2) [i.e., $dU = dQ = (3/2)nR\,dT$] that C_V is constant:

$$C_V(T, V) = \frac{3}{2}nR \qquad \text{[ideal gas]} \quad (9.4)$$

Here, and throughout this book, we assume that ideal gases consist of point particles, unless explicitly stated otherwise.

Note that C_V is extensive—which is why it is large for a large heat bath, leading to negligible ΔT_{sur} for the example discussed at the start of this section.

⊳⊳⊳**Helpful Hint:** It is not C_V itself, but rather the intensive *molar* quantity, $C_{V,\text{m}} = C_V/n$, that is more useful in practice (e.g., that one will find in a reference table). The heat capacity per unit mass—or *specific heat*, $C_V/(nM)$—is a closely related quantity, preferred in engineering.

⊳⊳⊳***Don't* Try It!!** Don't assume that $\Delta U = Q = C_V \Delta T$, even at constant V! Though often used in calorimetry applications, this expression is only valid when C_V is independent of T, or when ΔT is small. More generally, one has $dU = dQ = C_V\,dT$, and the integrated form,

$$\Delta U = Q = \int_{T_i}^{T_f} C_V(T)\,dT \qquad \text{[const } V\text{]} \quad (9.5)$$

9.2 Enthalpy & Heat Capacity at Constant Pressure

In calorimetry—and in many other contexts, too—we find that T and V are a natural choice of independent variables to use with U. In principle of course, any other variable pair could also be used, but there is a definite preference for thinking of $U = U(T, V)$.

One very nice property of Equation (9.1) is that it effectively turns a path function (Q) into a state function (ΔU). Path functions—particularly W—can be difficult to measure in real applications; thus a great simplification results when W is known in advance to be zero, as in the case of bomb calorimetry.

On the other hand, most real processes do not occur under constant *volume* conditions, but rather, under constant *pressure* conditions. Unfortunately, there is no special relationship between ΔU and Q at constant P. We therefore seek a *new* thermodynamic quantity, X, such that:

A "test tube" experiment, for example, is open to the air, and therefore subject to a constant $P = P_{\text{sur}} = P^\circ = 1$ atm.

- X is an extensive state function with dimensions of energy.
- X is closely related to U.
- The natural variables for X are T and P.
- $\Delta X = Q$ at constant P.

We will discuss the idea of "natural variables" further, in Sec. 14.1.

Such a quantity, should it exist,would be very useful in practice—because for any constant P process, it would avoid entirely the need to worry about work. Such a quantity does indeed exist; it is called the *enthalpy*.

Would we be talking about it if it didn't exist?

Definition 9.2 *The* enthalpy, *denoted 'H', is defined to be*

$$H = U + PV \qquad \text{[always]} \quad (9.6)$$

For calorimetry and other purposes, it is natural to define a new heat capacity quantity from $H(T, P)$.

Definition 9.3 *The* heat capacity at constant pressure, *denoted 'C_P', is defined to be the partial derivative of $H(T, P)$ with respect to T at constant P—i.e.,*

$$C_P = \left(\frac{\partial H}{\partial T}\right)\bigg|_P \qquad \text{[always]} \quad (9.7)$$

▷▷▷ ***Don't* Try It!!** Don't *ever* use H in the form of Equation (9.6), when computing C_P using Equation (9.7)! You must first use the $V(T, P)$ equation of state to get rid of the V dependence in Equation (9.6).

▷▷▷ **Helpful Hint:** As per the Don't Try It on p. 67, the expression $\Delta H = Q = C_P\,\Delta T$ at constant P is only an approximation to:

$$\Delta H = Q = \int_{T_i}^{T_f} C_P(T)\,dT \qquad \text{[const } P\text{]} \quad (9.8)$$

HIGHLIGHTS: Enthalpy

Proof: *Enthalpy has desired properties*

From Definition 9.2, H is an extensive state function quantity with dimensions of energy, closely related to U. Moreover, the last term in Equation (9.6) effectively transforms the variable V into P, as seen in the differential analysis below:

$$dH = \overbrace{dU}^{} + PdV + VdP$$

$$= dQ - PdV + PdV + VdP = dQ + VdP \tag{9.9}$$

$$\therefore \Delta H = Q \qquad \text{[const } P\text{]} \tag{9.10}$$

Note that an *infinitesimal* change, proceeding on the equation of state away from an initial state that is in equilibrium, is necessarily *reversible*. Thus,

$$dU = dQ + dW = dQ - P_{\text{sur}}\, dV = dQ - PdV.$$

Calculation: *Three ways to compute enthalpy change*

1. for constant P, from Q: $\Delta H = Q$ [Equation (9.10)].
2. for constant P, from ΔU and ΔV: $\Delta H = \Delta U + P\Delta V$ (see Helpful Hint on p. 74).
3. for any change: $\Delta H = (H_f - H_i)$ [because H is a state function; Equation (7.1)].

Comparison: *Enthalpy vs. internal energy*

- H is introduced merely for convenience (U is more fundamental).
- $H > U$ always (because $P > 0$ and $V > 0$).
- $C_P > C_V$ always (because V increases with T at constant P).
- $\gamma = (C_P/C_V)$ is key for reversible adiabatic change.

Example: *Ideal gas*

$$H(T, P) = U + PV = \frac{3}{2}nRT + nRT = \frac{5}{2}nRT$$

$$C_V = \frac{3}{2}nR \quad ; \quad C_P = \frac{5}{2}nR \quad ; \quad \gamma = (C_P/C_V) = \frac{5}{3}$$

$$\therefore H = \gamma\, U = \frac{5}{3}U > U$$

In practical terms, C_P is generally more useful than C_V; for example, C_P is what is usually meant when the term "heat capacity" is used by itself (i.e., with no additional qualifier). That said, the relation between the two heat capacities has important thermodynamic relevance. The heat capacity ratio, $\gamma = C_P/C_V$, plays a key role in reversible adiabatic change (γ is also called the *adiabat coefficient*). For the ideal gas, γ is constant; however, its value depends on whether or not point particles are presumed (see Sections 13.1 and 16.2).

Enthalpy does not even satisfy a conservation law, for example.

So what *is* enthalpy, really? Enthalpy is by no means a *true* energy, and is therefore far less fundamentally important than the internal energy. Probably the best way to think of enthalpy is simply as a practical tool—an energy-*like* quantity, better suited to constant P processes than is U itself. Working with H at constant P, Q effectively becomes a state function, and W can be ignored altogether. This is certainly handy, however...we would probably never have a need for enthalpy, were it not for the fact that we happen to like performing experiments at constant pressure.

9.3 Other Partial Derivative Quantities

From the Texts: Multivariable calculus not your strong suit? Many of the reference texts have good "refresher" sections that you can consult—along with Sec. 9.4 of this book.

According to the rules of calculus, the *total differential* for the state function $U(T, V)$ is given by:

$$dU = \left(\frac{\partial U}{\partial T}\right)\bigg|_V dT + \left(\frac{\partial U}{\partial V}\right)\bigg|_T dV \tag{9.11}$$

The first partial derivative above is C_V. The second partial derivative is also a useful quantity; it has dimensions of pressure, and is called the *internal pressure*, denoted 'π_T':

$$\pi_T = \left(\frac{\partial U}{\partial V}\right)\bigg|_T \qquad \text{[internal pressure]} \tag{9.12}$$

π_T can be measured using a *Joule apparatus*.

The internal pressure describes how U changes with V at constant T. Rather like Z (p. 30), π_T thus describes *intermolecular interactions* (Section 5.3). Note that $\pi_T = 0$ for the ideal gas; see Section 15.3 for additional discussion.

Other important partial derivative quantities can be obtained from the total differential for the $V(T, P)$ equation of state:

$$dV = \left(\frac{\partial V}{\partial T}\right)\bigg|_P dT + \left(\frac{\partial V}{\partial P}\right)\bigg|_T dP \tag{9.13}$$

The two partial derivatives in Equation (9.13) measure how V changes with respect to T and to P, respectively. These are thus quite useful in engineering, in the characterization of materials—but only after they are first converted into intensive quantities, via division by V.

We could just as well divide by n to obtain *molar* quantities, but this is not done in practice.

In this manner, we obtain the following "relative expansion quantities":

$$\alpha = \left(\frac{1}{V}\right)\left(\frac{\partial V}{\partial T}\right)\bigg|_P \qquad \text{[expansion coefficient]} \quad (9.14)$$

$$\kappa_T = -\left(\frac{1}{V}\right)\left(\frac{\partial V}{\partial P}\right)\bigg|_T \qquad \text{[isothermal compressibility]} \quad (9.15)$$

Note the minus sign in Equation (9.15); this is introduced so that κ_T itself is positive—since V always *decreases*, when P increases at constant T.

Why stop now? Why not continue to differentiate everything in sight, thereby creating a zillion new state function quantities? We could certainly do this, but there would be very little point. In practice, new thermodynamic quantities are introduced for one of three reasons:

1. They are theoretically meaningful.
2. They are experimentally useful.
3. They connect other quantities that we care about, through partial derivative relations.

Reason 3 will become a bit clearer in the next section.

9.4 Partial Derivatives & Differentials

We have made many references to "differential forms," "total differentials," and "partial derivatives." These are the building blocks of multivariable calculus, and also, useful tools for understanding thermodynamic change. Accordingly, in this section we explore some of the underlying mathematics in greater detail.

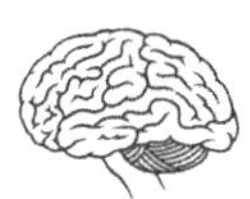

▷▷▷ **To Ponder...** Warning! As with Chapter 6, the material in this section may be somewhat advanced and/or not so immediately relevant for some students. Accordingly, some readers may prefer to proceed directly to Chapter 10.

Consider a generic mathematical function $f(x, y)$, in the two independent variables, x and y. We understand the basic idea that the "derivative" represents how f changes with respect to the independent variable—but since there are *two* of these instead of one, there are two independent ways in which the variables can change, and therefore two partial derivatives.

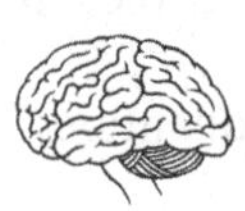

▷▷▷ **To Ponder...** In fact, there are *many* ways that the variables can change—corresponding to many diagonal paths in the (x, y) "state space" (one such path is indicated in the marginal figure on this page)—and thus, there are many different partial derivatives.

Only two of these many partial derivatives are *independent*, however. It is convenient to take these as:

$\left(\frac{\partial f}{\partial x}\right)\Big|_y$ associated with the horizontal path, and dx

$\left(\frac{\partial f}{\partial y}\right)\Big|_x$ associated with the vertical path, and dy

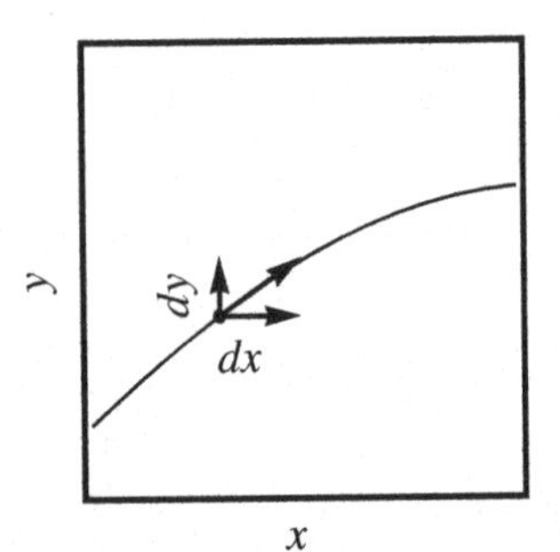

For the horizontal path, the infinitesimal change in f is given by $df = (\partial f/\partial x)|_y\, dx$. For *any* path, it is given by the the *total differential*:

$$df = \left(\frac{\partial f}{\partial x}\right)\Big|_y dx + \left(\frac{\partial f}{\partial y}\right)\Big|_x dy \tag{9.16}$$

Equation (9.16) is obtained in the infinitesimal limit of an *arbitrary* change—i.e., for any diagonal path. Diagonal paths correspond to directions along which some *third* variable—e.g., $z = z(x, y)$—is held fixed. As a consequence, for example, $(\partial f/\partial x)|_z \neq (\partial f/\partial x)|_y$.

▷▷▷ **Helpful Hint:** For any term in a total differential, the 'd' and lower '∂' coordinates must match [e.g., x, for the first term of Equation (9.16)]. Checking that this is so will help you to avoid making errors when working on problems.

Another useful feature of total differentials is the following cross-derivative "trick." Suppose we encounter a total differential dg, for some quantity g, that may—*or may not*—be a state function. In general, the total differential will take the form:

Students are often asked to go through this exercise for an f that is *already known* to be a state function; we feel that this is pointless.

$$dg = A(x, y)\, dx + B(x, y)\, dy \tag{9.17}$$

According to calculus, the order in which multiple partial derivatives are taken does not matter. Therefore, by comparing Equations (9.16) and (9.17), we find that g is a state function if and only if

$$\left(\frac{\partial A}{\partial y}\right)\Big|_x = \left(\frac{\partial B}{\partial x}\right)\Big|_y \qquad \text{[exact differential]} \tag{9.18}$$

In this case, dg is called an *exact differential*.

MATHEMATICAL DISCUSSION: Partial Derivatives

$\left(\frac{\partial f}{\partial x}\right)\Big|_y$ = partial derivative of $f(x, y)$ with respect to x, keeping y fixed.

$\left(\frac{\partial f}{\partial y}\right)\Big|_x$ = partial derivative of $f(x, y)$ with respect to y, keeping x fixed.

The specification of a partial derivative depends *crucially* on what is held fixed—*just as much* as on what is being varied! Thus, $(\partial f/\partial x)$ is meaningless, and $(\partial f/\partial x)|_y \neq (\partial f/\partial x)|_z$. Note that in the latter derivative, the variables have been changed so that $f = f(x, z)$—thus, $(\partial f/\partial x)|_z$ corresponds to a *diagonal*, rather than horizontal, path in (x, y) space.

Question: *What does it mean when a politician says "all other factors being equal, a budget increase for program X will lead to outcome Y"?*

It means absolutely nothing. He or she would have to explicitly list "all other factors," in order for such a statement to make any sense—and of course, no one is ever actually able to do this. **Science Doesn't Care.**

Example: *Consider the function,* $f(x, y) = x^2y^2$.

We obtain $(\partial f/\partial x)|_y = 2xy^2$, by treating y as a constant, and taking the usual derivative of f with respect to x. To compute $(\partial f/\partial x)|_z$, we must first *transform* f *to the new variables,* (x, z), by getting rid of all y's via the substitution $y = y(x, z)$. Suppose $z = xy^2$, so that $y(x, z) = \sqrt{z/x}$. Then, $f(x, z) = xz$, and $(\partial f/\partial x)|_z = z = xy^2$. Thus, $(\partial f/\partial x)|_y \neq (\partial f/\partial x)|_z$.

▹▹▹ **Try It!!** The classic example of an *inexact* differential in thermodynamics is the reversible infinitesimal work, $dW = 0\,dP - P dV$. You can easily verify that dW does *not* satisfy Equation (9.18).

From the Texts: Some authors prefer the term *"perfect differential."*

▹▹▹ **Helpful Hint:** As per the second Helpful Hint and the Don't Try It on p. 56, every term in a differential expression such as Equation (9.16) must have one 'd' factor. Don't combine X, ΔX, and dX in the same expression!

⊳⊳⊳**Helpful Hint:** On the other hand, you can always convert an 'X' expression into a 'Δ' (or 'd') expression, simply by introducing a 'Δ' in front of every term. Moreover, any constants or fixed variables can be "pulled out" of the Δ's. Examples: Equations (7.2), (8.1), and (9.2).

Entropy

"I often say that when you can measure what you are speaking about, and express it in numbers, you know something about it; but when you cannot measure it, when you cannot express it in numbers, your knowledge is of a meagre and unsatisfactory kind; it may be the beginning of knowledge, but you have scarcely in your thoughts advanced to the state of Science, whatever the matter may be."

—Lord Kelvin

"These go to eleven."

—Nigel Tufnel, *Spıñal Tap*

A Conceptual Guide to Thermodynamics, First Edition. Bill Poirier.

Companion website: http://www.conceptualthermo.com

Chapter 10

Entropy & Information Theory

Anderson: Sec. 4.4, Sec. 4.15; **Atkins:** Sec. 3.2, Sec. 15.1, Sec. 15.4; **Atkins-life:** Sec. 2.2, Sec. 2.4; **Baierlein:** Sec. 2.4, Sec. 2.7; **Callen:** Sec. 1-10, Sec. 15-1, Sec. 16-9, Sec. 17-1; **Cengel:** Sec. 7–6, Sec. 7–7, pp. 465–468; **Chang:** Sec. 4.2, pp. 98–100; **Elliott:** Sec. 1.3, pp. 129–136, Sec. 4.3; **Engel:** p. 86, Sec. 5.3, p. 102, Sec. 30.1, Sec. 32.4; **Faure:** Sec. 11.6; **Kittel:** pp. 29–36, pp. 39–44, p. 50, pp. 445–447; **Levine:** Sec. 3.3, Sec. 3.7; **McQuarrie:** Sec. 20–5; **Moran:** Sec. 6.1, Sec. 6.3, Sec. 6.8.2, Sec. 6.11; **Prausnitz:** Appendix B; **Reif:** Sec. 2 · 4, Sec. 2 · 5, Sec. 3 · 3, p. 111, p. 115, Sec. 3 · 10, Sec. 3 · 11; **Sandler:** Sec. 4.1; **Schroeder:** Sec. 2.1, Sec. 2.6, Sec. 3.2; **Silbey:** Sec. 3.2, Sec. 3.6, p. 571; **Smith:** Sec. 5.4, Sec. 5.11, Sec. 16.3; **Tinoco:** p. 55, p. 60, pp. 62–65, pp. 162–163

10.1 Why Does Entropy Seem So Complicated?

Students and authors alike dread the subject of *entropy*. To be sure, entropy, denoted 'S', is a state function like no other. Equally clear is that it is of central importance to thermodynamics—being intimately associated with two of the four Laws. Yet many individuals—even science professionals who use entropy every day—wrestle with the basic question of what it actually *is*, exactly.

It is an interesting exercise to review how the different authors address this question. Among the reference textbooks, entropy—or its explanation—has been described (sometimes apologetically!) as follows: "mysterious;" "uncomfortable;" "the most misunderstood property of matter;" "less than satisfying;" "less than rigorous;" "confused;" "rather abstract;" "rather subjective;" "difficult to fully comprehend…physical

From the Texts: See if you can find these descriptions in *your* primary textbook [**Try It!!**].

A Conceptual Guide to Thermodynamics, First Edition. Bill Poirier.

Companion website: http://www.conceptualthermo.com

implications that are hard to grasp;" "not a household word, like energy;" etc. Perhaps **Cengel** puts it most poignantly and directly:

From the Texts: Y. A. **Cengel** and M. A. Boles, *Thermodynamics: An Engineering* Approach, p. 345. (Reproduced with permission, copyright © 2012 by McGraw-Hill Education.)

This does not mean that we know entropy well, because we do not. In fact, we cannot even give an adequate answer to the question, "what is entropy?"

It does not have to be this way.

First and foremost, entropy is a *quantity*. This means that any description of entropy that purports to be a *definition* had better have a number attached to it. Moreover, the physical meaning of that number should be clear—e.g., we must know precisely what it means to "add one" to it, in the same way that we know precisely what it means to add one calorie of heat to a gram of water.

Recall our "definition of a definition," Definition 1.1, on p. 4.

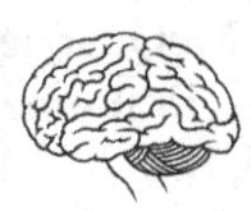

▷▷▷ **To Ponder**... We laugh at a certain fictional rocker, for putting his faith in a particular amplifier because it goes "one louder." But we scientists are really no better, if we cannot clearly explain just exactly what it means to go "one more entropic." Can you do this? Can your professor? If not, don't worry; you will by the end of this chapter.

Try It !!

As entropy *is* a quantity, the usual descriptions of it as a measure of molecular "disorder," "randomness," "chaos," etc.—though conceptually useful—are clearly not sufficient. Next in the hierarchy of less-than-fully-satisfactory explanations is the statement that entropy is "the quantity that increases under a spontaneous irreversible change" (Section 12.3)—which, of course, still does not really tell us what it is.

The most standard definition of entropy is the oldest one, known as the *thermodynamic* definition. It stems from a purely macroscopic approach that does not rely on a molecular description at all. This definition is a bit "funny," in that it is usually expressed in differential form—specifically, via a relation involving the temperature and the *reversible heat*, or heat absorbed during a reversible change:

From the Texts: Some authors refer to this as the *macroscopic* or *classical* definition.

From the Texts: Many authors use 'Q_{rev}' to denote the reversible heat.

$$dQ = TdS \qquad \text{[reversible, infinitesimal]} \qquad (10.1)$$

The thermodynamic definition has the huge advantage of being quantitative. However, it, too, has some significant limitations.

The first limitation—obvious, though not always acknowledged—is that it technically applies only to *reversible* changes. The usual workaround is to prove that dS is an exact differential and therefore path-independent. Nevertheless, given that the most important application of entropy—the Second Law (see Chapter 12)—concerns only *irreversible* processes, it seems

This limitation is explicit in Eq. (10.1), a form which we prefer to the standard Q_{rev} construction (see the Helpful Hint on p. 94).

strange that the definition itself should be restricted to *reversible* processes only!

The second limitation of the thermodynamic definition is that—being a differential—it is only capable of predicting entropy *changes*, ΔS, and not absolute entropy *values*, S. This distinction is an important one that we will encounter often, most notably in the Third Law (see Section 13.3).

The two limitations already imply that the thermodynamic definition is not a *true* definition, in the sense of Definition 1.1.

However, by far the most widely acknowledged limitation of the thermodynamic definition of entropy is that, in and of itself, it provides very little physical understanding. It is presumably for this reason that even advanced textbooks still retain some qualitative "disorder"-type descriptions, in addition to Equation (10.1). Evidently for this reason, also, it has in recent years become fashionable for textbooks to include a brief blurb on the *statistical definition* of entropy. This approach—developed by Boltzmann, and exemplified in his famous formula [Equation (10.2), p. 84]—offers a molecular origin of entropy, and is one of the most rigorous.

fashionable, at least, in chemistry and chemical engineering...

On the other hand, at first glance the Boltzmann approach seems to have nothing to do with thermodynamic entropy. It is, in fact, quite a challenge to quantitatively relate the two definitions using straightforward means; one traditionally resorts to complicated statistical mechanical proofs that all but the advanced physics texts would rather avoid. As **Raff** puts it,

> *...we lack the tools required to rigorously demonstrate the connection between the [two types of] entropy...*

From the Texts: L. M. **Raff**, *Principles of Physical Chemistry, Part A* (Prentice-Hall, 2001), p. 170.

In this book, we adopt the *information theory* approach to entropy, whose advantages in this regard will become clear as we go along.

10.2 Entropy as Unknown Molecular Information

Without further ado, here is a qualitative version of the information definition of entropy:

Definition 10.1 (qualitative) Entropy *is the amount of molecular information that a macroscopic observer* does not *know about the system.*

Recall from Section 3.1 that thermodynamics provides a complete description of reality at the *macroscopic* scale only; at the *molecular* scale, it is largely incomplete. Thus, precise knowledge of the thermodynamic state of the system [e.g., of the thermodynamic variables (T, V)] tells us very little about the molecular state. The details of the latter—i.e., positions, velocities, and internal states of every constituent molecule—must therefore be considered information that is mostly unknown to a *macroscopic observer*. According to Definition 10.1, the *amount* of this missing molecular information is the entropy.

i.e., us...

Some comments are in order.

or the "pineal gland" if you prefer...

First, note that entropy is the *fundamental link* between the macroscopic and molecular descriptions of nature—the "*medulla oblongata,*" if you will—between thermodynamics and statistical mechanics, as per Section 2.3. Thus, rather than favoring either the macroscopic or molecular viewpoint, it behooves us to recognize at the outset that both are necessary in order to achieve a full understanding of this quantity.

Second, Definition 10.1 has some very interesting ramifications for the Second Law of Thermodynamics. However, a full discussion on this point will have to wait until Section 12.4.

These are molecular states of the *whole system*, not of a single molecule. (see the Don't Try It on p. 21).

Third, note that even the qualitative Definition 10.1 already suffices to prove that *entropy is a thermodynamic state function*. The set of molecular states associated with a given thermodynamic state clearly depends only on the thermodynamic state itself. Moreover, the entropy value evidently varies from one thermodynamic state to another—i.e., $S = S(T, V)$ is not constant. This can be understood as follows.

reasonable, and also correct; the two are not always the same...

Recall from Figure 3.1 that there are many possible molecular states associated with each thermodynamic state. It is reasonable to assume that some thermodynamic states have *more* molecular states associated with them than do others. Presumably, these thermodynamic states are characterized by more unknown molecular information—and therefore, greater entropy. Intuitively, we expect larger T and V to correspond to a larger number of possible molecular states, and thus, to higher S values. This is, in fact, almost always the case.

It should be mentioned that the state function aspect of entropy is something that other treatments—particularly those based on the thermodynamic definition of Equation (10.1)—must go to a fair amount of trouble to validate. The usual proof has several stages, invoking the rather complex Carnot cycle (which we also consider, in Section 13.2, but for different reasons). Here, the desired result is seen to follow immediately from Definition 10.1, without any additional effort.

The equivalence with statistical entropy will turn out to be obvious.

On the other hand, in order to demonstrate that the *information entropy* as defined here is indeed the same as the usual thermodynamic entropy, we are obliged to prove that the former satisfies Equation (10.1). Curiously, though a number of the reference textbooks now include *both* the thermodynamic definition of Equation (10.1) *and* the statistical (quantitative information) definition of Equation (10.2), in most of these, no real attempt is made to prove the equivalence of the two. We will do so in Section 11.5 of this book, for the special case of the ideal gas.

From the Texts: Many authors do mention statistical entropy, but they do not directly connect this to thermodynamic entropy.

10.3 Amount of Information

Nigel Tufnel would probably be OK with it, though...

So far, our information definition of entropy is only qualitative—and therefore, according to Lord Kelvin, unscientific (see quote on p. 75)—because we have not yet specified exactly what is meant by "amount of information." Fortunately, the field of *information theory* provides us with exactly the quantitative formula that we need.

a branch of probability and statistics...

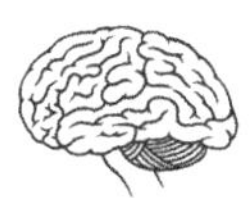

▷▷▷ **To Ponder...** In fact, it is the same basic formula that is used in computer science—to determine, e.g., how much information can be stored on your computer, or how many bytes are needed to store the MP3 file for your favorite song.

The framework is that of a hypothetical observer, who gains information by performing a measurement that yields one outcome from a number of possibilities.

Definition 10.2 (quantitative) *The* amount of information *gained from any measurement is the logarithm of Ω, where Ω is the number of possible outcomes.*

Definition 10.2 is technically only valid if the possible outcomes are all equally likely; if not, we need a means of estimating the "effective" number of possible or available outcomes—e.g., by ignoring those that are extremely unlikely.

It makes intuitive sense that the amount of information should increase with Ω. Why is it that $\log(\Omega)$ is the amount of information, though, and not simply Ω itself? *This is so that the amount of information gained from separate, independent measurements is additive*. By "additive," we mean that the total amount of information gained from independent measurements is just the *sum* of the amounts of information gained from each measurement separately.

exactly why will become clear in the upcoming dice examples...

Some instructive examples may be found in the rolling of dice. In the boxes below, six-sided dice are presumed, and each die roll is considered to be independent of (i.e., unrelated to, or unaffected by) the others.

Note that "dice" is the plural; the singular form is "die."

EXAMPLE: Rolling One Die

A single, six-sided die is thrown. An observer then looks at the die to see what number was rolled.

Question: *How much information is gained from the measurement?*

The die roll "experiment" has $\Omega = 6$ possible outcomes. By Definition 10.2, the amount of information gained is therefore $\log(6)$.

EXAMPLE: Rolling Two Dice

Now two dice are thrown—each a different color, so that the observer can tell them apart. After both dice are thrown, the observer looks to see what number was rolled on each of the two dice.

Question: *How much information is gained now?*

First, we must determine the total number of possible outcomes. Since each die has six possibilities, and any combination of these may be observed, the total number of possible outcomes for two dice is:

$$\Omega = 6 \times 6 = 6^2 = 36$$

(We assume the dice are not "loaded," and that the two rolls are independent. Also, the *distinguishability* of the two dice, due to their different colors, is very important.) The amount of information gained is:

$$\log(6 \times 6) = \log(6^2) = 2\log(6)$$

Note that the amount of information gained from rolling two dice is exactly *twice* that of a single die roll, which makes good intuitive sense. The extension to N dice is straightforward:

$$\begin{aligned}\langle\text{number of possible outcomes}\rangle &= \Omega = 6^N \\ \langle\text{amount of information}\rangle &= \log(\Omega) = N\log(6)\end{aligned}$$

Thus, information is indeed additive—thanks to the logarithm function (see the Log Blog post on page 83).

COMPARISON: Dice vs. Molecules

Dice	Molecules
N dice	N molecules
independent die rolls	noninteracting free particles (ideal gas)
6 possible outcomes per die roll	Ω_1 available states for each molecule

The amount of information is technically a dimensionless quantity.

In general, the numerical value of the amount of information depends on the base of the logarithm, which is left unspecified in Definition 10.2 (p. 81). Different bases correspond to different "units" of information (see Log Blog post and Correspondence on p. 83).

LOG BLOG: Latest Post (see also earlier post on p. 63).

▷▷▷ **Helpful Hint:** The most important property of logarithms, especially for information theory, is that they effectively convert *multiplication* into *addition*. Thus: $\log(xy) = \log(x) + \log(y)$; $\log(x^n) = n\log(x)$.

▷▷▷ **Helpful Hint:** Logarithms also convert *division* into *subtraction*: $\log(x/y) = \log(x) - \log(y)$. Thus, $\log(V_f/V_i) = \log(V_f) - \log(V_i) = \Delta(\log V)$. Note also:

- If $x > 1$, then $\log(x) > 0$.
- If $x < 1$, then $\log(x) < 0$.
- If $V_f > V_i$ (true expansion), then $(V_f/V_i) > 1$, so $\log(V_f/V_i) > 0$.
- If $V_f < V_i$ (compression), then $(V_f/V_i) < 1$, so $\log(V_f/V_i) < 0$.

This is how we wind up with expressions like Eq. (8.7)

▷▷▷ **Helpful Hint:** The integral, $\int f(x)\,dx$, of every *power law* function, $f(x) = x^p$, is another power law function—*except* when $p = -1$, in which case $\int(1/x)\,dx = \ln(x)$.

This fact *also* explains Eq. (8.7).

▷▷▷ **Helpful Hint:** The most general form of the logarithm is $\log_b(x)$, where b is the *base*. The logarithm base can be changed according to the formula, $\log_{b_{new}}(x) = \log_{b_{old}}(x)/\log_{b_{old}}(b_{new})$, which effectively changes the *unit* of information.

CORRESPONDENCE: Log Base and Unit of Information

Log Base (b)	Unit of Information
2	bit
Euler constant, $e \approx 2.71828$	natural log (ln) unit
10	common log (log) unit
$2^8 = 256$	byte

10.4 Application to Thermodynamics

Combining the qualitative Definition 10.1 (p. 79) with the information theory results from Section 10.3, we are now in a position to provide a *quantitative* information definition of the entropy for a thermodynamic system, as follows:

Definition 10.3 (quantitative) *The* entropy, *denoted 'S', is defined to be*

$$S = k\ \ln(\Omega), \qquad \text{[always]} \quad (10.2)$$

where 'ln' denotes the natural log, Ω is the number of possible or available molecular states (of the whole system) associated with the thermodynamic state, and k is the Boltzmann constant.

Equation 10.2 is the famous *Boltzmann entropy formula*—i.e., the statistical definition developed by Boltzmann and by Gibbs.

although their original interpretation and application are somewhat different than the information theory approach adopted here...

In principle, computing the entropy for a given thermodynamic state reduces to simply counting the number of corresponding molecular states, Ω. However, this is not at all easy to do in practice. For one thing, Ω is absolutely *enormous*—much, *much* larger than Avogadro's number, N_A, and in fact much closer to $\exp(N_A)$. The logarithm in Equation (10.2) helps a lot, by reducing the mind-bogglingly huge Ω down to something that is "merely" on the order of N_A. The factor of k then converts this molecular-scale number to macroscopic units, leaving us in the end with typically quite reasonable SI values for S.

This is true for macroscopic systems, for which N is on the order of N_A.

Nevertheless, there is still the problem of actually *counting* Ω. This is essentially impossible in the non-ideal case when N is on the order of N_A; we must therefore make do with the largest N values that can be achieved using the latest computers, and hope that these are "large enough" to obtain reasonable results.

or we must make approximations; **Science Doesn't Care**...

However, in the special case of the ideal gas, Ω *can* be computed exactly, by exploiting the fact that all molecules behave independently. In Chapter 11, the thermodynamic state function for S—i.e., the explicit function, $S = S(T, V)$—is derived for the ideal gas from Equation (10.2). In addition, the corresponding expression for the entropy change, ΔS, matches exactly the standard result obtained by integrating the thermodynamic differential form of Equation (10.1).

Remember the dice examples from pp. 81–82. The molecular independence stems from the lack of intermolecular interactions (Sec. 5.3).

Before moving on, we briefly address the issue of dimensions. As per the second Helpful Hint in the Log Blog post on p. 63, the argument of the logarithm in Equation (10.2)—i.e., Ω itself—is dimensionless. The amount of information is *also* dimensionless, although the factor of k that appears in Equation (10.2) results ultimately in S dimensions of energy over temperature (J/K, in SI units). As discussed in Section 4.4 however, k is simply a conversion factor between temperature and energy, introduced essentially just for convenience. Some authors therefore prefer to regard the dimensionless quantity, $\sigma = \ln(\Omega) = S/k$, as the "fundamental entropy." The

Much of the rest of this section is fairly advanced and/or esoteric, though all readers are invited to read the last To Ponder of the chapter (p. 85).

From the Texts: Kittel, pp. 42–44.

conjugate variable—i.e., $\tau = kT$, is then called the "fundamental temperature." Note that τ has dimensions of energy.

When working with σ and τ, the contrived kelvin unit no longer plays a direct role. Such an approach—already eminently sensible, even without the information definition of entropy—becomes highly compelling when the latter is adopted, given that σ itself now becomes the amount of information as directly measured in natural log units. In any event, the J/K units of S can perhaps best be regarded as an arbitrary convention, or historical accident.

Conjugate variables will be introduced in Sec. 14.1. Note that this σ has nothing to do with standard deviation.

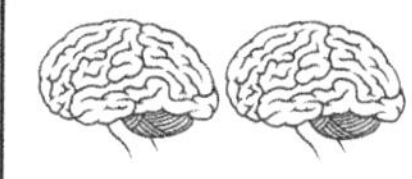

▷▷▷ **To Ponder...*at a deeper level.*** Even so, some authors have turned this situation on its head, reasoning that the information definition of entropy cannot be physically correct, *because* it leads naturally to a dimensionless quantity. Entropy—it is argued—is physically related to energy, and must therefore evidently possess energy-like dimensions. The first part of this claim is undeniably true. As for the second part...would anyone argue that heat capacity and specific heat—or molality and molarity—must be unrelated because they have different dimensions? In reality, the only dimensional requirement here is that the temperature-entropy conjugate variable product must have dimensions of energy—which *is* true, whether one uses TS *or* $\tau\sigma$.

From the Texts

For convenience, we work with the *conventional* quantities, T and S—even though these are less natural for the information approach than would be an analysis based on τ and σ.

Unit issues aside, what would be most compelling and useful, from both the practical and pedagogical standpoints, is a direct, quantitative connection between the thermodynamic and information definitions of entropy—say, a simple derivation of Equation (10.1), directly from Equation (10.2), which does not explicitly require the heavy machinery of statistical mechanics such as configurations, ensembles, and the like. Such a connection will be provided in Chapter 11, for the special case of the ideal gas.

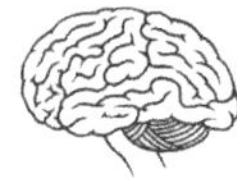

▷▷▷ **To Ponder...** We conclude this chapter by reconsidering the question posed at the beginning: what does it really mean to go "one more entropic"? Adopting the fundamental entropy σ, and the "bit" (p. 83) as the unit of information, this simply means doubling the number of molecular states available to the whole system.

Chapter 11

Entropy & Ideal Gas

Anderson: Sec. 4.11 Sec. 7.5.1, Sec. 13.2; **Atkins:** Sec. 3.2, Sec. 3.3, Chap. 15; **Atkins-life:** Sec. 2.4; **Baierlein:** Sec. 2.1, Sec. 2.5, Sec. 2.6, pp. 104–108; **Callen:** Sec. 16-1, Sec. 16-2, Sec. 16-10, Sec. 17-2; **Cengel:** Sec. 7–7, Sec. 7–9, pp. 701–705; **Chang:** p. 86; **Elliott:** pp. 136–138, Sec. 4.3; **Engel:** Sec. 5.3, Sec. 5.4, Sec. 5.5, Sec. 5.9, Sec. 30.1, p. 778, Sec. 31.2, Sec. 31.4, p. 828, Sec. 32.4; **Faure:** n/a; **Kittel:** pp. 29–36, p. 61, p. 72–80; **Levine:** Sec. 3.4, Sec. 4.5; **McQuarrie:** Sec. 20–3, Sec. 20–6, Sec. 20–9; **Moran:** Sec 6.5; **Prausnitz:** p. 45, Appendix B; **Reif:** Sec. 2 · 5, Sec. 3 · 6, p. 111, Sec. 4 · 5, p. 215, Sec. 7 · 1, Sec. 7 · 2, Sec. 7 · 3; **Sandler:** Sec. 4.4; **Schroeder:** Sec. 2.5, pp. 77–81; **Silbey:** Sec. 3.2, pp. 84–85, Sec. 3.6, pp. 570–573, p. 575, Sec. 16.3; **Smith:** Sec. 5.4, Sec. 5.5, Sec. 16.3, p. 660; **Tinoco:** pp. 163–164

11.1 Measuring Our Molecular Ignorance

Our goals in this chapter are two-fold. First, we seek to prove that starting from the statistical, or information definition of S as presented in Equation (10.2), we can derive the thermodynamic form of Equation (10.1), under reversible conditions. As discussed in Chapter 10, the general proof is too advanced for the scope of this book. For the ideal gas, on the other hand, a simple and straightforward derivation is possible. For simplicity, we assume an ideal gas of point particles—although the derivation can easily be generalized to incorporate rotating molecules or those with other internal structure, provided that they are noninteracting.

From the Texts: Interested readers should consult sources such as **Engel**, **Kittel**, and **Reif**.

and/or molecules subjected to external forces...

The second goal is to use Equation (10.2) to derive an expression for the entropy state function, S, as an explicit function of the thermodynamic variables, (T, V), for the ideal gas. We verify that the result is equivalent to the famous *Sackur-Tetrode equation* [Equation (11.11), p. 93] to within an *additive constant*. This again confirms the agreement of the two entropies, for the ideal gas case. In this fashion, we establish a connection between the statistical and thermodynamic definitions of entropy, using the information theory approach.

i.e., equivalent apart from a "shift," $S \to S+D$, where D is the additive constant.

A Conceptual Guide to Thermodynamics, First Edition. Bill Poirier.

Companion website: http://www.conceptualthermo.com

▷▷▷ **Helpful Hint:** When working with S as a state function, the natural thermodynamic variables are T and V, for reasons described below. See Section 9.2 for a discussion of natural variables, and Section 14.1 for a more formal discussion of "conjugate variables."

It all boils down to counting states properly—i.e., to the calculation of Ω. For an ideal gas of point particles, the molecular state of the *whole system* is specified via the $6N$ molecular coordinates, (x_i, y_i, z_i) and $(v_{x,i}, v_{y,i}, v_{z,i})$. As macroscopic observers, we do have some *a priori* information about the values of these $6N$ coordinates, but our knowledge is incomplete. Our lack of knowledge about the particle positions is directly associated with the system volume, V. Likewise, our lack of knowledge about the particle velocities (and kinetic energies) is associated with the temperature, T. These two sources of "ignorance" both contribute to the total entropy of a given thermodynamic state, (T, V), and can be treated separately.

See Sec. 3.3, and specifically the Don't Try It on p. 21.

In this context, *a priori* means "without making a measurement of the molecular state."

As discussed in Section 10.4, Ω is enormous—even *infinite*, in a sense, as we will see. For non-ideal systems, intermolecular interactions make the calculation of Ω difficult. The calculation becomes much simpler in the ideal gas case because the individual particles are statistically independent. Thus, as per Section 10.3, Ω can be obtained directly from Ω_1 (the number of states available to a single particle) via

Remember that the independence stems from the lack of intermolecular interactions (Sec. 5.3).

$$\Omega = \Omega_1^N. \qquad \text{[ideal gas]} \quad (11.1)$$

It remains only to determine Ω_1, in terms of the available single-particle positions, (x, y, z), and velocities, (v_x, v_y, v_z)—where the 'i' subscripts are henceforth dropped (see Section 6.2).

11.2 Volume Contribution to Entropy

Consider the change in entropy, ΔS, brought about via isothermal (true) expansion of the ideal gas. Because T is fixed, there is no change in the (statistically-averaged) particle velocities. Therefore, all of ΔS is due to the increase in volume, which in turn is associated with changes in the particle positions.

Now consider the physical box within which the system resides, as depicted in Figure 11.1. As macroscopic observers, we know that every particle must lie somewhere *inside* the box, but we have no idea precisely *where*. Thus, if a measurement of a given particle's position were to be performed, we know *a priori* that there would be *zero* probability of finding the particle outside the box, and equal probability of finding it at any point inside the box.

The system is *closed*, as per Sec. 4.1.

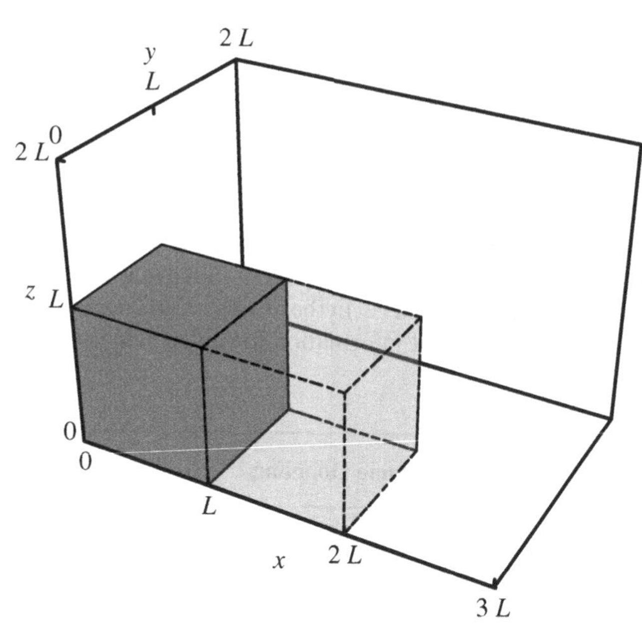

Figure 11.1 Expansion of a cubical box. Cubical box of volume $V = L^3$ (opaque gray, solid lines), denoting physical limits of the system. Without performing an explicit measurement of precise particle positions, we know that such a measurement would find every particle to be *inside* rather than outside, of the cube. This knowledge provides partial *a priori* information about the molecular state. Under expansion of the box out to $x = 2L$, (translucent gray, dashed lines), our knowledge of the particle positions *decreases*, and the entropy accordingly *increases*.

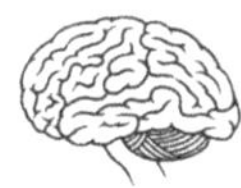

▷▷▷ **To Ponder...** Although it is convenient to think of entropy as the amount of information gained by the observer as a result of precise particle measurement, it is important to realize that such a measurement is imagined to be hypothetical only. The reality is that modern science actually *does* enable us to peer at individual molecules, in a manner that would have been inconceivable to the founding fathers of thermodynamics. In that sense, we are no longer true *macroscopic* observers.

It is natural to think of each (x, y, z) point as constituting a different "position state" for a single particle. The total number of position states available to a given particle is thus equal to the number of points inside the box. Since space is continuous, this number is technically infinite. However, that is not too problematic, since (as we will see) the determination of ΔS requires only the *relative* number of available states. In any case, it is evident that the number of points inside the box—and therefore Ω_1 (at constant T)—is proportional to the box volume, V. We thus have

$$\Omega_1 = AV, \tag{11.2}$$

where A remains constant under isothermal expansion. In principle, A can depend on T, but it must be independent of V.

From Equations (10.2), (11.1), and (11.2), we thus find

$$S = k \ln\left(\Omega_1^N\right) = Nk \ln(\Omega_1) = Nk \ln(AV). \quad \text{[ideal gas]} \quad (11.3)$$

S is also extensive in the *non*-ideal case, provided N is sufficiently large.

For macroscopic systems, $\Omega = (AV)^N$ grows enormously quickly with V. However, the logarithm converts the exponent into a linear N factor. Thus, *entropy is extensive*, as it should be. In the ideal gas case, the extensivity of S follows from the additivity of information for independent measurements (Section 10.3).

EXAMPLE: Isothermal Volume Doubling

Suppose that the box that contains the system is expanded to twice its original size (as in Figure 11.1), under isothermal conditions. We thus have $(V_i = V) \rightarrow (V_f = 2V)$.

Question: *How do the corresponding values of* Ω_1, Ω, *and S change?*

- for a *single particle*: the number of states is doubled—i.e., $\Omega_1 \rightarrow 2\Omega_1$.
- for the *whole system*: $\Omega_i = \Omega_1^N \rightarrow \Omega_f = 2^N\Omega_1^N = 2^N\Omega_i$. For macroscopic N, Ω_f is *absolutely huge* compared to Ω_i.
- for the *entropy*:

$$\begin{aligned} \Delta S = S_f - S_i &= k \ln(\Omega_f) - k \ln(\Omega_i) \\ &= k \ln\left(\frac{\Omega_f}{\Omega_i}\right) = k \ln(2^N) \\ &= Nk \ln(2) \end{aligned} \quad (11.4)$$

Equation (11.4) is a quantitative prediction, with no unspecified parameters, for the change in entropy of an ideal gas under isothermal doubling of the volume. Note that in the final expression for ΔS, the actual *values* of Ω_i and Ω_f do not appear. In general, all that matters for entropy *change* is the *ratio*, (Ω_f/Ω_i). Note also that $\Delta S > 0$, as expected.

See the Log Blog post on p. 83.

More generally, if the volume change is arbitrary (i.e., from any V_i to any V_f), but still isothermal,

$$\Delta S = S_f - S_i = Nk \ln(AV_f) - Nk \ln(AV_i) = Nk \ln\left(\frac{V_f}{V_i}\right).$$

$$\text{[ideal gas, const } T] \quad (11.5)$$

Again, the actual volume *values* do not matter for ΔS; only the ratio of final to initial volumes is relevant. Nor does the unspecified constant A enter into the final form for ΔS. Consequently, a specific numerical prediction can be made for the ΔS value for any isothermal gas expansion.

Remember, though, that A is only "constant" with respect to V; the T dependence will be explored next.

11.3 Temperature Contribution to Entropy

Thus far, we have considered only changes in V at constant T. To round out the full thermodynamic story, we must also consider changes in T at constant V. Because V is fixed, there is no change in our information about the molecular positions; these therefore play no role in the origin of ΔS in this case. The entropy does change, however—owing to a change in the available *velocity* states, induced by the change in temperature.

▷▷▷**Helpful Hint:** Remember: V corresponds to the available molecular *positions*; T corresponds to the available molecular *velocities*.

Recall from Section 6.2 that for a system in thermal equilibrium, the distribution of particle velocities is not uniform, but adheres to the Maxwell-Boltzmann distribution [Equation (6.2) and Figure 6.1]. Recall, also, that this same distribution describes all three velocity components of all system particles—although for simplicity, we refer to a single, 'v_x' component only.

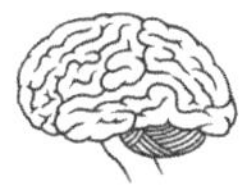

▷▷▷ **To Ponder...** Why can't we use the Maxwell distribution of speeds [Equation (6.6)] instead of the Maxwell-Boltzmann distribution? Because it is based on *speeds* v, rather than velocities (v_x, v_y, v_z), and therefore does not incorporate complete molecular state information [some of the velocity information is lost in Equation (6.6)—namely, the directional information].

For a given velocity component of a given particle, we need to somehow estimate the number of available states associated with the Equation (6.2) distribution. The way that we do this is by defining an "effective" range of allowed v_x values. We can then use this range as a measure of the number of available v_x states—just like we did in Section 11.2 for the position states. In both cases, the bigger the (effective) range, the greater the number of available states.

e.g., 'v_x'...

From Section 6.2, we learn that the standard deviation, σ_{v_x}, is a good measure of the effective range of the v_x distribution. For the Maxwell-Boltzmann distribution, in particular, the standard deviation—and hence, the number of available v_x states—is found to increase proportionally with the square root of the temperature (i.e., $\sigma_{v_x} \propto T^{1/2}$ [Equation (6.5)]). Since there are three velocity components for each particle, each of which is independent of the other two, the total number of (v_x, v_y, v_z) velocity states available to each particle must be proportional to $T^{1/2} \times T^{1/2} \times T^{1/2} = T^{3/2}$.

In this manner, we are led to

$$\Omega_1 = BT^{3/2}, \tag{11.6}$$

where the "constant" B is independent of T, but may depend on V. The first part of Equation (11.3) then yields

$$S = k \ln\left(\Omega_1^N\right) = Nk \ln(\Omega_1) = Nk \ln\left(BT^{3/2}\right). \quad \text{[ideal gas]} \tag{11.7}$$

Like the last part of Equation (11.3), the new form of the ideal gas entropy as given by Equation (11.7) also depends on an unspecified quantity—in this case B. Once again, however, an unambiguous prediction can be obtained for ΔS (now under constant volume conditions) with no unspecified parameters:

$$\Delta S = Nk \ln\left(BT_f^{3/2}\right) - Nk \ln\left(BT_i^{3/2}\right) = \frac{3}{2} Nk \ln\left(\frac{T_f}{T_i}\right)$$
$$\text{[ideal gas, const } V\text{]} \tag{11.8}$$

11.4 Combined Entropy Expression

Remember that $A = A(T)$ and $B = B(V)$.

Comparing the rightmost parts of Equations (11.3) and (11.7), we find that $A(T)V = T^{3/2}B(V)$. This can only be true if $A(T) = CT^{3/2}$ and $B(V) = CV$, where C is a true *thermodynamic constant*. This means that C depends neither on T nor V, and is therefore independent of the thermodynamic state. We thus obtain

$$S(T, V) = Nk \ln\left(CVT^{3/2}\right), \quad \text{[ideal gas]} \tag{11.9}$$

i.e., the desired explicit $S(T, V)$ state function for the ideal gas.

Depending on your background, the next bit of discussion on quantum mechanics might be extraneous, and could be skipped.

Note that in Equation (11.9) above, the value of the constant C is still unspecified. This means that the above analysis is insufficient to provide us with absolute entropy values—i.e., specific numerical values for $S(T, V)$ itself—though it does specify $S(T, V)$ to within an additive constant. Nevertheless, C *is indeed* found to have a specific value in nature, which can be determined using *quantum mechanics* (a discipline that lies outside the scope of this book).

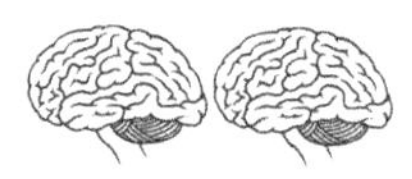

▷▷▷ **To Ponder...*at a deeper level.*** As discussed, classical theory technically predicts an *infinite* value for C, because the number of available states, Ω, is infinite. According to quantum theory, on the other hand, the *uncertainty principle* effectively "quantizes" the continuous classical molecular states, thus replacing them with a finite number of discrete quantum states. In this way, the "infinite states" difficulty of Section 11.2 is avoided.

For reference, we provide the quantum mechanical value of C; it is

$$C = \left(\frac{2\pi mke}{h^2}\right)^{3/2}, \qquad (11.10)$$

where e is the Euler constant, and h is Planck's constant. Substitution of Equation (11.10) into Equation (11.9) then leads exactly (after incorporating a correction due to the indistinguishability of quantum particles) to the *Sackur-Tetrode equation*:

$$S(T, V) = Nk \ln\left[\frac{V(2\pi mkT)^{3/2}}{h^3 N}\right] + \frac{5}{2}Nk \quad \text{[ideal gas]} \quad (11.11)$$

▷▷▷ **Helpful Hint:** Whenever h or $\hbar$ appears in a formula, you know that quantum mechanics is somehow involved.

▷▷▷**Helpful Hint:** Some authors write the Sackur-Tetrode equation in a form that—unlike Equation (11.11), and the second Helpful Hint in the Log Blog post on p. 63—does *not* use dimensionless logarithm arguments. This situation arises because the logarithm contribution is split into two or more separate terms. When using such a form to solve problems, you must be *extremely* careful with units.

Although C remains unspecified in a purely classical analysis, its value is immaterial insofar as entropy *change* is concerned. This is because C—being a multiplicative factor within the logarithm argument of Equation (11.9)—is converted into an additive constant, which must cancel out

We are now talking about completely arbitrary changes of thermodynamic state—i.e., both T and V may vary.

in any calculation of ΔS. This can be seen more explicitly as follows:

$$\begin{aligned}\Delta S &= Nk\ln\left(CV_f T_f^{3/2}\right) - Nk\ln\left(CV_i T_i^{3/2}\right)\\ &= Nk\ln\left(\frac{CV_f T_f^{3/2}}{CV_i T_i^{3/2}}\right)\\ &= Nk\ln\left(\frac{V_f T_f^{3/2}}{V_i T_i^{3/2}}\right)\end{aligned} \tag{11.12}$$

Due to a cancellation of factors, C does not appear in the final expression above. A more suggestive form of this same equation—valid for arbitrary thermodynamic changes—is provided below:

$$\Delta S = Nk\ln\left(\frac{V_f}{V_i}\right) + \frac{3}{2}Nk\ln\left(\frac{T_f}{T_i}\right) \qquad \text{[ideal gas]} \tag{11.13}$$

Note from Equation (11.13) that ΔS separates cleanly into additive "volume entropy" and "temperature entropy" contributions, each of which involves a logarithm with a dimensionless argument. These two contributions are associated, respectively, with changes in our knowledge about the particle positions and velocities. Note also that Equation (11.13) reduces to Equation (11.5) in the special case of constant T changes, and to Equation (11.8) for constant V changes.

When solving problems involving the entropy of the ideal gas, remember to use the 'ln' key on your calculator, as per the first Helpful Hint on p. 63.

11.5 Entropy, Heat, & Reversible Adiabatic Expansion

We now derive the well-known relation between entropy and heat, under reversible infinitesimal changes of state—i.e., Equation (10.1). Although this differential relation is true even for non-ideal systems, our proof encompasses only the ideal gas case.

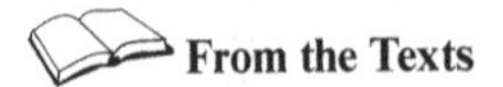

▷▷▷**Helpful Hint:** You should be aware that there is a tendency among many authors to express the heat absorbed under reversible conditions as if it were its own special quantity—the *reversible heat*, Q_{rev}. We do not regard this to be a thermodynamic quantity in its own right; in reality, this is just the usual heat Q, but restricted to reversible thermodynamic processes only. Similar comments apply to *reversible work* (Section 8.4, p. 60).

From Equations (9.16) and (11.9), the total differential dS for the state function $S(T, V)$ is

$$dS = \left(\frac{\partial S}{\partial T}\right)\bigg|_V dT + \left(\frac{\partial S}{\partial V}\right)\bigg|_T dV$$
$$= \frac{3}{2}\left(\frac{Nk}{T}\right)dT + \left(\frac{Nk}{V}\right)dV \qquad (11.14)$$

Consider the following derivation:

DERIVATION: Ideal Gas of Point Particles

$$\frac{3}{2}Nk\,dT = dU \qquad \text{[Equation (5.7) and Helpful Hint on p. 74]}$$

$$\frac{Nk}{V} = \frac{nR}{V} = \frac{P}{T} \qquad \text{[Equation (4.5)]}$$

$$\therefore\ dS = \frac{dU}{T} + \frac{P\,dV}{T} = \frac{1}{T}(dU + P\,dV) \qquad (11.15)$$

Now, since $PdV = -dW$ under reversible conditions, and $dQ = dU - dW$, the quantity in parentheses in Equation (11.15) is clearly dQ, under reversible conditions. Therefore,

$$dS = \frac{dQ}{T}, \qquad \text{[reversible, infinitesimal]} \quad (11.16)$$

which is identical to Equation (10.1).

The relation between heat and entropy has important ramifications for *reversible adiabatic expansions* (see Chapters 13 and 16). For most thermodynamic quantities, this is the hardest special case because it involves changes in *all three* thermodynamic variables, T, P, and V. A student might therefore be well inclined to *panic* at the prospect of having to work out how a tricky quantity such as entropy might change under a reversible adiabatic expansion! In fact, though, this calculation could not be easier; under these conditions, $dQ = 0$, and so Equation (11.16) implies that $\Delta S = 0$. This is a general result, true for non-ideal as well as ideal gases, so **It's OK to be Lazy**.

and also for the Second Law (see Chap. 12)...

The information ramifications of ΔS being zero under a reversible adiabatic change are interesting to consider. Under these conditions, Ω remains constant, and thus the total amount of information that we have about the system remains unchanged. On the other hand, this does *not* imply that the actual molecular states that are available to the system *themselves* remain the same. Instead, Equation (11.13) implies that the increase in volume

If that were true, then there would be no change of thermodynamic state!

entropy must be exactly balanced by a corresponding *decrease* in temperature entropy.

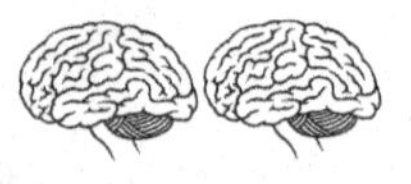

▷▷▷ **To Ponder...*at a deeper level.*** In effect, we *gain* information about the particle velocities, while simultaneously *losing* information about their positions, in a kind of "thermodynamic uncertainty principle."

Chapter 12

Second Law & Spontaneous Irreversible Change

Anderson: Sec. 4.2, Sec. 4.4.3, Sec. 4.12; **Atkins:** pp. 94–96, Sec. 3.2(e), Sec. 3.5(c); **Atkins-life:** pp. 69–71; **Baierlein:** Sec. 2.2, Sec. 2.3, Sec. 2.5, Sec. 2.8; **Callen:** Sec. 4-1, Sec. 4-5; **Cengel:** pp. 277–280, p. 285, pp. 290–292, Sec. 6–6; **Chang:** pp. 81–83, Sec. 4.3; **Elliott:** pp. 98–100, Sec. 4.5; **Engel:** Sec. 5.1, Sec. 5.2, Sec. 5.5, Sec. 5.6, Sec. 5.7; **Faure:** Sec. 11.6; **Kittel:** pp. 45–51, p. 240; **Levine:** Sec. 3.1, Sec. 3.5, Sec. 3.8, Sec. 4.2, Sec. 4.9; **McQuarrie:** pp. 816–821, Sec. 20–4; **Moran:** Sec. 5.1, Sec. 5.2, Sec. 5.3, Sec. 5.11; **Prausnitz:** n/a; **Reif:** Sec. 3 · 11, Sec. 5 · 12; **Sandler:** pp. 98–107; **Schroeder:** Sec. 2.3, Sec. 2.6, Sec. 3.4; **Silbey:** Sec. 3.2, Sec. 3.4; **Smith:** pp. 159–160, Sec. 5.6, Sec. 5.8; **Tinoco:** pp. 13–14, pp. 55–56, pp. 60–62

12.1 Heat Engines & Thermodynamic Cycles

Perhaps the best way to introduce the Second Law of Thermodynamics is still the original way—i.e., in the context of heat engines. Conceptually, the notion of a thermodynamic *heat engine* is straightforward: it is a device that absorbs heat energy, (some of) which it then converts into work done on its surroundings.

Implicit in the use of the word “engine” is the idea that the device can be used repeatedly over time. Thus, whereas some kind of thermodynamic change is obviously required in order to generate nonzero W, we cannot rely on changes that *perpetually* increase V. Instead, heat engines use *cycles*—loop paths in state space that return to where they started, thereby incorporating both expansion *and compression* stages.

See also Sec. 13.2 and Fig. 13.2, and the box on p. 56.

Similar considerations must also hold for *temperature*; any T increase must be balanced by an equal and opposite T *decrease*, somewhere else within the cycle. A naive individual might imagine that Q must also be balanced around such a cycle—whatever heat is absorbed must later be released. This is *false*; $Q \neq 0$ around a cycle, precisely because Q is not a state function.

certainly not a sophisticated student of thermodynamics such as yourself...

A Conceptual Guide to Thermodynamics, First Edition. Bill Poirier.

Companion website: http://www.conceptualthermo.com

Science Doesn't Care...

The work must also be nonzero around a cycle—otherwise, heat engines would not function! Therefore—and although it can make solving thermodynamics problems a pain—the path-function aspect of Q and W is actually critically important. Note finally that the cycle must loop around *clockwise* in (P, V) space, in order for $W < 0$. Thus, the expansion stage(s) occurs at higher P (and higher T) than the compression stage(s). We will examine heat engines more quantitatively in Section 13.2, in the context of the Carnot cycle; here, the discussion remains more conceptual.

12.2 Traditional Statements of the Second Law

Total W, like total Q, has "work in" and "work out" stages—though in practice, these are not usually considered separately.

The "naive observer" from Section 12.1 has now learned, presumably to his or her amazement, that heat engines *do* exist. Around a cycle, the total Q—i.e., the difference between absorbed and released heat—is *not* zero, but is in fact equal to the total work done, $-W$. Note that the bigger the heat difference, the greater the work performed—motivating the design of heat engines for which released heat is as small as possible.

naïveté often *does* seem to flit from one extreme view to its opposite...

Is it possible to build a "perfect" heat engine—i.e., one for which there is *no* released heat, so that all absorbed heat is converted into useful work? Let's suppose that our naive observer now imagines this new scenario to be the case—i.e., the exact *opposite* of his or her earlier stance. He or she would, once again, be wrong. The reason turns out to be the Second Law.

From the Texts: **Atkins**; **Cengel**, Sec. 6–2 and p. 304; **Elliott**; **Engel**, p. 89; **Kittel**, p. 49; **Levine**, p. 79; **Moran**; **Prausnitz**; **Reif**; **Sandler**, p. 105; **Smith**, p. 160.

Most engineering and many science textbooks introduce the Second Law through a standard series of *Statements*—more or less identical to the conclusion of the previous paragraph, and to each other. More precisely: the basic description above is the *Kelvin Statement*; the (necessary!) restriction to cycles (introduced by Max Planck) gives rise to the *Kelvin-Planck Statement*; the converse statement (i.e. running the cycle in reverse) is the *Clausius Statement*.

Textbooks that present the formal Statements above may do so for several reasons—e.g., because of historical tradition, and/or because heat engines are still very relevant today. Another likely reason, though, is because they typically introduce the Second Law *before* introducing entropy. Even those science texts that do not list the formal Statements almost invariably provide an "arrow" diagram, showing heat flow in being split into work flow out plus heat flow out. Such a diagram constitutes a pictorial representation of the same basic principle.

From the Texts: This obligatory figure is not reproduced here, but can almost certainly be found in your primary text [**Try It !!**].

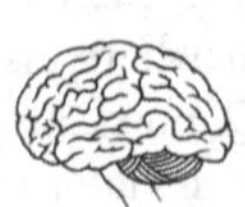

⊳⊳⊳ **To Ponder...** The idea of heat flowing *into* the engine, rather than out of it, may seem strange to modern-day automobile drivers—but not to their turn-of-the-last-century counterparts! (There were, indeed, steam-powered automobiles ... See Appendix C.) This is, in fact, one of the primary differences between the steam engine and the internal combustion engine.

The Second Law Statements above preclude the existence of what is called a *perpetual motion machine of the second kind.* Proposals for machines of this kind have enjoyed a rich and colorful history—leading to innumerable patent applications, and even out and out fraud. All such devices are impossible, however—due to the fact that the Second Law is, in fact, a *law* (Section 2.2). Indeed, as useful as the heat engine application turns out to be, it is merely one example of a much broader and more important general principle—the Second Law—whose real significance depends fundamentally on the concept of entropy.

From the Texts: Some entertaining discussion may be found in: **Cengel**, p. 293; L. M. **Raff**, *Principles of Physical Chemistry, Part A* (Prentice-Hall, 2001), p. 145.

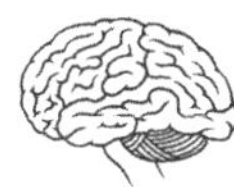

▷▷▷ **To Ponder...** Rudolf Clausius (see Appendix A), of the above Clausius Statement, discovered the entropy interpretation of the Second Law, and also coined the term "entropy" from a Greek word meaning "transformation." In doing so, he very fortunately abandoned his earlier term for this fundamental quantity—which was "*verwandlungsinhalt*"!

12.3 Entropy Statement of the Second Law

In the same manner that energy plays a fundamental role vis-à-vis the First Law, entropy plays a fundamental role in the Second Law, given here:

Second Law (total system): *The entropy of the total system* increases *under any* spontaneous *irreversible thermodynamic change:*

$$\Delta S_{\text{tot}} = \Delta S + \Delta S_{\text{sur}} > 0 \qquad \text{[irreversible]} \qquad (12.1)$$

We emphasize from the start that in its most fundamental form above, the Second Law of Thermodynamics is a statement about the *total system.* In that respect, Equation (12.1) very much resembles the most fundamental form of the First Law—i.e., Equation (7.2).

From the Texts: Equivalently, but with less generality, some authors prefer to frame the Second Law in terms of an *isolated system* (Sec. 7.1).

One critical difference from Equation (7.2), however, is that Equation (12.1) is an *inequality*, rather than an equality (conservation law). Much can and has been made of this inequality feature of the Second Law. Some authors—e.g., **Raff**—put a negative spin on this:

> *[The Second Law]...is the only scientific law that is stated in negative terms. It tells us what we cannot do.*

From the Texts: L. M. **Raff**, *Principles of Physical Chemistry, Part A* (Prentice-Hall, 2001), p. 144.

Some people may like to believe that the "sky is the limit"—that we can achieve anything at all, if we are merely ingenious enough. But generally speaking, this expectation is unrealistic and naive. There are *always* limits, and it is in fact a very *good* thing to know those limits in advance. In any case, there *are* a few other "inequality limit" laws in science, such as the speed-of-light limit on the speed of moving particles (special relativity) and also the absolute zero limit on temperature (intimately connected with the Third Law, Section 13.3).

because of the Second Law, the US Patent and Trademark Office no longer accepts applications for perpetual motion machines...

In any case, the most important ramification of the Second Law inequality is that *spontaneous irreversible change occurs in one direction only.* Consider running any spontaneous irreversible process in reverse; ΔS_{tot} is now negative, and thus forbidden by the Second Law. Such "nonspontaneous" irreversible processes, though hypothetically possible, do not actually occur in nature. Many examples from everyday experience may be considered—a glass knocked off a counter falls on the ground and shatters into many pieces, but the reverse process is never seen.

As kids, my brother and I were amazed by the evil deeds that Superman would do—when running our Super 8 mm reel of the classic 1978 movie in reverse...

From the Texts: A truly excellent discussion of the philosophical ramifications of the Second Law may be found in **Levine**, Sec. 3.8 (and also Sec. 4.9).

The above, "arrow of time" feature of the Second Law is all the more profound, given that the underlying molecular laws of physics themselves do *not* appear to have any time directionality built into them. It is an inherently "emergent" property (Section 2.1) that continues to occupy the great scientific minds of the present day, such as Stephen Hawking.

Entropy and the Second Law are often interpreted in energetic terms. More precisely, entropy relates to the *distribution* or *dispersal* of energy—with the Second Law implying that over time, energy becomes more uniformly distributed throughout the total system. According to the Second Law, everything in the universe will eventually reach thermal equilibrium—making the operation of heat engines impossible, for instance. Whether this depressing "heat death" will ever actually be realized is still an open question—though it has preoccupied people since the late 19th century (see Appendix B). In any event, energy dispersal, etc., ultimately pertains to the available molecular states—thus leading us to consider the information theory viewpoint.

taking the total system to be the entire universe...

Science Doesn't Care...

12.4 Information Statement of the Second Law

Definition 10.1 (p. 79) has quite interesting ramifications for the Second Law of Thermodynamics, which can be interpreted in information terms as follows:

Second Law (information): *Information about the total system decreases under a spontaneous irreversible change.*

In the Statement above, and henceforth, the word "information" is used as a shorthand for:

Similarly, "we/us" shall often represent the macroscopic observer of Sec. 10.2.

amount of molecular information known by a macroscopic observer

Since the Second Law implies an increase in the total system entropy as an inevitable consequence of each spontaneous irreversible change that takes place, our ignorance about the universe increases—and therefore our information *decreases*—as time progresses. This rather depressing conclusion may seem counterintuitive. Keep in mind, however, that "information" in this context has a very limited and literal meaning, in terms of the positions, velocities, etc. of molecules.

particularly to students, whose goal is presumably to *increase*, rather than decrease, their knowledge about the world...

Put another way, the number of molecular states available to the total system increases over time, during the course of a spontaneous irreversible change. Why should this be the case? This is because of the causative sudden external change—which always serves to remove some previous macroscopic restriction on the total system (a tire is suddenly punctured; a fixed wall is suddenly released; heat is suddenly allowed to flow, etc.). Removing macroscopic restrictions increases the number of available molecular states—thereby placing the total system in a "less special" final thermodynamic state.

Note that the sudden change is "external" to the system, but internal to the *total system*...

Ultimately, the Second Law holds because the universe itself started out in a very special thermodynamic state—for which there were many restrictions, few corresponding molecular states, and a very low S_{tot}. By its demise, the universe may well see all of its macroscopic restrictions removed, and all of its parts in perfect equilibrium with each other—the aforementioned "heat death."

or "heaven," depending on your point of view (see quote on p. 9)...

Recalling the "arrow of time" discussion in Section 12.3, it is worth considering how the Second Law "emerges" in the large N limit—e.g., in the context of the punctured tire system example (box on p. 27). Consider monitoring a single air molecule, starting from the instant that the tire is punctured ($t = 0$). Within a certain time interval, we might well observe this molecule moving from the outside to the inside of the tire, and think nothing of it; this event may be only slightly less probable than its opposite (i.e., moving from inside to outside), and in any case, the molecular laws of physics are time-reversible.

From the Texts: See also **Chang**, pp. 82–83. Chang also has a great quote on p. 89: "Just thinking about entropy increases its value." (Reproduced with permission, copyright © 2005 by University Science Books.)

On the other hand, if we were to observe *all* of the $N \approx N_A$ particles of the macroscopic (total) system, moving from outside to inside the tire, during the same time interval, this would be an *extremely* improbable—even *unnatural*—event. Such a scenario accurately describes what would have to happen if the punctured tire "movie" were run in reverse.

Some authors prefer to speak of "natural" and "unnatural" processes, rather than "spontaneous" and "nonspontaneous" processes—and perhaps wisely so. Such nomenclature correctly suggests, e.g., that the spontaneous accumulation of air molecules inside the punctured tire—and the commensurate reduction in the number of available molecular states and S_{tot}—*is* in fact statistically possible. However, the likelihood of such a process actually occurring in nature becomes *overwhelmingly improbable* when N is large. Almost always, $S_{\text{tot}}(t)$ increases with t during the course of a spontaneous irreversible change, as indicated in Figure 12.1.

From the Texts: e.g., **Engel**, p. 85.

This would constitute a large-magnitude statistical fluctuation away from the curve in Fig. 12.1; see also Fig. 5.1 and subsequent discussion...

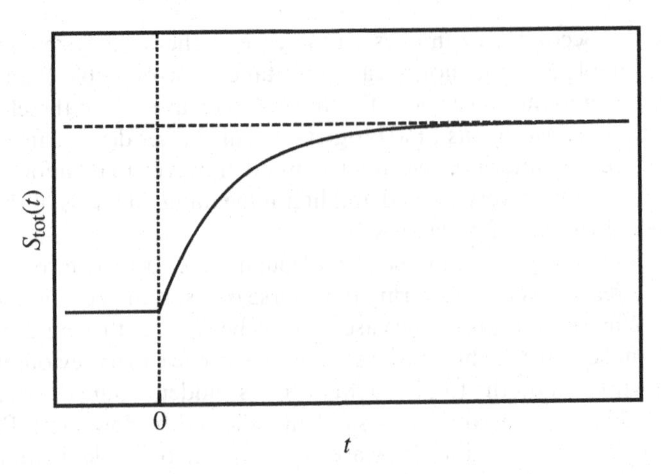

Figure 12.1 Entropy of the total system as a function of time. Entropy of the total system, S_{tot} (solid curve), as a function of time, t, during the course of a spontaneous irreversible change—in response to a sudden external change applied at $t = 0$ (vertical dotted line). Initially, $S_{tot}(t)$ increases with t, as the total system explores more of the molecular states now available to it. Eventually, $S_{tot}(t)$ approaches a new constant (horizontal dashed line), indicating that a new equilibrium has been reached.

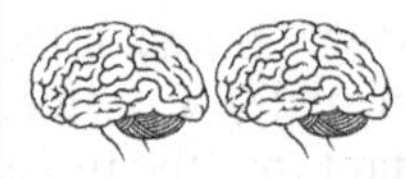

▷▷▷ **To Ponder...*at a deeper level.*** Since entropy is a state function, and the thermodynamic state may not even be well defined during the course of an irreversible thermodynamic change (Section 8.4), how can we determine $S_{tot}(t)$ in Figure 12.1? This can be done by monitoring the particle *probability distributions* (Section 6.1) over time. When the system is out of equilibrium, the position distribution is *not* uniform, and the velocity distribution is *not* the Maxwell-Boltzmann distribution. Only for $t < 0$ and $t \rightarrow \infty$—i.e., the initial and final (equilibrium) states, respectively—do the distributions take these standard forms.

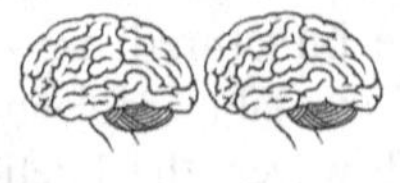

▷▷▷ **To Ponder...*at a deeper level.*** Though extremely rare when N is large, statistical fluctuations that appreciably reduce the value of S_{tot} *do* occur—it is merely a matter of waiting long enough. Boltzmann himself considered the role that such fluctuations might play in a "heat death" end-of-the-universe scenario. Recently, this question has once again become a hot topic in the field of cosmology, where fluctuations are (seriously) imagined to give birth to whole new universes. (No one can accuse cosmologists of not taking ideas to their absolute extreme limits—it is their *raison d'être*, after all...) On the other hand, rather than creating a whole new universe comprised of a multitude of stars, galaxies, intelligent observers, etc., would it not be far easier simply to create a single brain that merely (and erroneously) *believes* in all of this rich external structure? Both scenarios are consistent with experience—yet the latter, as improbable as it sounds, is presumably far more likely. This is the *Boltzmann brain paradox*.

12.5 Maximum Entropy & the Clausius Inequality

Figure 12.1 implies that after a sudden external change, equilibrium is restored again only after $S_{\rm tot}$ reaches a maximum value. This is known as the *principle of maximum entropy*, and it is very useful for predicting the final equilibrium state that ensues after a spontaneous irreversible change has occurred.

EXAMPLE: Two Subsystems in Mechanical Contact

The two subsystems A and B from the marginal figure on p. 27 are suddenly brought into mechanical contact. Assume that $T = T_A = T_B = \text{const}$, but that the initial pressures $P_{A,i} \neq P_{B,i}$, so that a spontaneous irreversible change occurs. Specifically, the movable wall moves until equilibrium is restored.

Question: *Assuming an ideal gas, what are the final subsystem states?*

- Because $V = V_A + V_B = \text{const}$ throughout the process, only V_A is independent. We can therefore write $S(V_A) = S_A(V_A) + S_B(V_A)$.
- The final state value, $V_A = V_{A,f}$, is that for which $S(V_A)$ is maximized. According to calculus, this is obtained by setting $dS/dV_A = 0$:

$$S(V_A) = N_A k \ln(AV_A) + N_B k \ln(A[V - V_A]) \quad \text{[Eq. (11.3)]}$$

$$\frac{dS}{dV_A} = kA\left(\frac{N_A}{V_A}\right) - kA\left(\frac{N_B}{[V - V_A]}\right) = 0 \quad \text{[maximum entropy]}$$

$$\left(\frac{N_A}{V_{A,f}}\right) = \left(\frac{N_B}{V_{B,f}}\right) \quad \text{[final volume result]}$$

$$\therefore P_{A,f} = P_{B,f} \quad \text{[multiply by } kT\text{; ideal gas law]}$$

Note: 'N_A' is the number of particles in subsystem A, whereas '$N_{\rm A}$' is Avogadro's number.

and also the state of *minimum information*, or "maximum ignorance"...

Thus, the *maximum entropy* state for the above example is also the *mechanical equilibrium* state, as per Section 4.3. Similar arguments for two subsystems in *thermal* contact can be used to show that—in that case—heat flows until S is maximized, which occurs when $T_{A,f} = T_{B,f}$ (i.e., the condition for thermal equilibrium).

for an *isolated system*...

In the box above, we used the subsystems picture of Section 4.3. Going forward, however, the total system or "system-plus-surroundings" picture will be more appropriate, because we will be deriving a new form of the Second Law that applies directly to the system itself.

▷▷▷***Don't* Try It!!** Don't assume that $\Delta S > 0$ for a spontaneous irreversible change! ΔS can also be *negative*—for the same reason that ΔU can be nonzero—because the First Law [Equation (7.2)] and the Second Law [Equation (12.1)] apply to the *total system*, not to the system itself.

but only by following a *different* path back (To Ponder on p. 61)...

Thus, whereas it is possible to restore the system to its original state, after it has first undergone an irreversible change, it is *never* possible to restore the *total system* to *its* original state, after an irreversible change. To do so would violate the Second Law.

An "irreversible" change is thus effectively reversible, insofar as the surroundings are concerned.

Building on the discussion in Section 7.1, a key difference between the system and surroundings is that the latter are regarded to be in equilibrium with themselves *throughout* an irreversible change. Thus, quantities like P_{sur} and S_{sur} are always well defined.

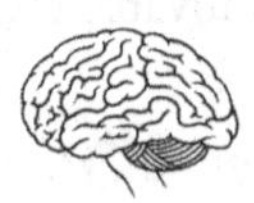

▷▷▷ **To Ponder...** Right around this point in many of the reference textbooks, you will notice that the 'ex' or similar subscript used to denote the surroundings, suddenly changes to something like 'sur'. Could this be to avoid the awkwardness of 'S_{ex}'? The serious point here: there is essentially *no difference* between the "externals" and the surroundings.

Try It!! Check to see if *your* textbook does this.

Consider the derivation provided in the box below. This gives rise to the *Clausius inequality*, a form of the Second Law that applies directly to the system:

Second Law (differential form): *Under any spontaneous irreversible infinitesimal thermodynamic change,*

$$dS > \frac{dQ}{T} \qquad \text{[irreversible, infinitesimal]} \quad (12.2)$$

Equation (12.2) above is thus analogous to the differential form of the First Law [Equation (8.1)]. Note that dS can indeed be negative—but only if dQ is also negative.

DERIVATION: Clausius Inequality

$$dS_{\text{tot}} = dS + dS_{\text{sur}} \quad \text{[differential form of Eq. (12.1)]}$$

$$dS_{\text{tot}} = dS + \frac{dQ_{\text{sur}}}{T_{\text{sur}}} \quad \text{[surroundings form of Eq. (11.16)]}$$

$$dQ_{\text{sur}} = -dQ \quad \text{[differential form of Eq. (7.4)]}$$

$$T = T_{\text{sur}} \quad \text{[system and surroundings in thermal equilibrium]}$$

$$\therefore dS_{\text{tot}} = dS - \frac{dQ}{T} > 0 \quad \text{[irreversible, Second Law]}$$

Chapter 13

Third Law, Carnot Cycle, & Absolute Entropy

Anderson: Sec. 4.16, Sec. 5.4, Sec. 5.5; **Atkins:** Sec. 3.2(c), Sec. 3.4; **Atkins-life:** Sec. 2.3; **Baierlein:** Sec. 3.1, Sec. 3.2, Sec. 3.7; **Callen:** p. 30, Sec. 4-6, Sec. 4-7, Sec. 4-8; **Cengel:** Sec. 6–3, Sec. 6–4, Sec. 6–7, Sec. 6–8, Sec, 6–10, p. 346; **Chang:** pp. 87–88, Sec. 4.5; **Elliott:** Sec. 3.1, Sec. 5.1, Sec. 9.5; **Engel:** Sec. 5.2, Sec. 5.8, Sec. 5.11; **Faure:** p. 162; **Kittel:** p. 49, pp. 228–240; **Levine:** Sec. 3.2, Sec. 5.7; **McQuarrie:** Sec. 19–4, Sec. 19–5, Sec. 20–6, Sec. 20–7, Sec. 21–4; **Moran:** Sec. 5.5, Sec. 5.6, Sec. 5.7, Sec. 5.9, Sec. 5.10, Sec. 13.5; **Prausnitz:** n/a; **Reif:** pp. 119–121, Sec. 4 · 6, Sec. 5 · 11; **Sandler:** Sec. 4.3, Sec. 6.8; **Schroeder:** pp. 25–26, Sec. 3.2, Sec. 4.1; **Silbey:** Sec. 3.3, Sec. 3.8, Sec. 3.9; **Smith:** Sec. 5.2, Sec. 5.3, Sec. 5.10; **Tinoco:** p. 14, pp. 56–60, pp. 66–69

13.1 Entropy & Reversible Change

The Second Law tells us what happens to S_{tot} under a spontaneous irreversible change. What happens to S_{tot} under a *reversible* change? Recall that *all* thermodynamic changes are "reversible" for the surroundings; therefore, a reversible change for the system must also be reversible for the total system.

Engineers distinguish *externally* from *internally* reversible processes.

Consequently, under a reversible change:

- ΔS_{tot} cannot be *greater* than zero, or the process would be spontaneous irreversible.
- ΔS_{tot} cannot be *less* than zero, or the process would be forbidden by the Second Law.
- $\therefore\ \Delta S_{\text{tot}} = \Delta S + \Delta S_{\text{sur}} = 0.$ [reversible]

A Conceptual Guide to Thermodynamics, First Edition. Bill Poirier.

Companion website: http://www.conceptualthermo.com

Recall the "information" shorthand from p. 100.

described at the end of Sec. 11.5...

assuming a true $\Delta V > 0$ expansion (see Helpful Hint on p. 51)...

From the Texts: This form is called "polytropic" (e.g., **Moran**, Sec. 3.15); general adiabats are called "isentropic" (e.g., **Cengel**, Sec. 7.4).

Thus, the information about the total system is *conserved* under a reversible change; whatever information is *gained* about the system is *lost* about the surroundings, or vice-versa. This situation is reminiscent of the balance between volume and temperature information that characterizes the system under reversible adiabatic expansions, resulting in $\Delta S = 0$. We now see that reversible adiabatic changes also satisfy $\Delta S_{sur} = \Delta S_{tot} = 0$. Expansions of this kind form a key part of the Carnot cycle (Section 13.2), and will therefore be reviewed here.

Reversible adiabatic expansion of ideal gas: Recall from Sections 8.5 and 9.1 that a reversible *isothermal* expansion can be achieved by placing a piston-cylinder apparatus in a heat bath, and then slowly easing up on P_{sur}. The resultant reversible path is an isotherm, given for the ideal gas by $P(V) \propto V^{-1}$.

To achieve reversible *adiabatic* expansion, we surround the apparatus in some *thermally insulating* material, rather than a heat bath—but otherwise we proceed similarly. Because work is done but no heat is absorbed, U and T both *decrease*. The resultant *reversible adiabatic path*, or "adiabat," is thus different from the corresponding isotherm—although both paths lie on the equation of state, as indicated in Figure 13.1.

For the ideal gas, the projection of the reversible adiabatic path onto (P, V) space yields $P(V) \propto V^{-\gamma}$, where γ is the adiabat coefficient (p. 70). Recall that γ is a constant, with a value greater than one (because $C_P > C_V$). Consequently, the adiabat represents a power law—like the corresponding isotherm, but with a larger inverse power, resulting in a steeper decline. This

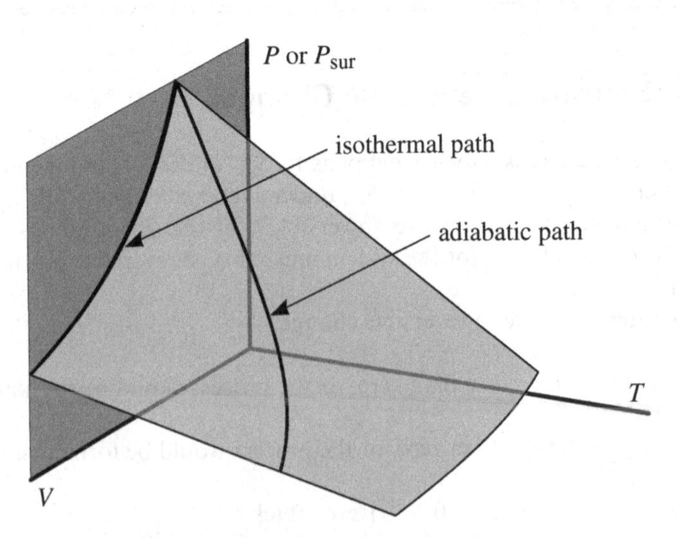

Figure 13.1 Reversible and isothermal paths. All *reversible* paths must lie on the *equation of state*—represented here as the translucent curved surface. All *isothermal* paths, including irreversible ones, must lie within a surface of constant T—such as the dark gray vertical plane indicated (and used also in Fig. 8.2). The intersection of these two surfaces defines the *reversible isothermal path*—but there are many other reversible paths along which T changes, such as the adiabatic path indicated.

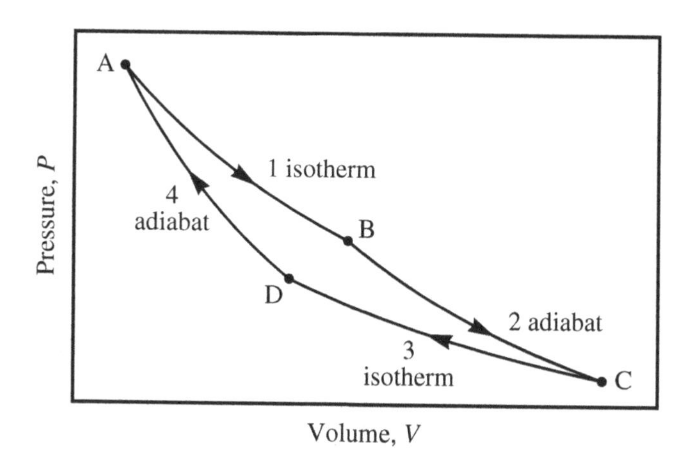

Figure 13.2 Four stages of the Carnot cycle. The four stages of the Carnot cycle, in terms of the piston-cylinder apparatus of Fig. 8.1. **Stage 1:** isothermal; insert apparatus in high $T = T_h$ bath, slowly expand. **Stage 2:** adiabatic; remove apparatus from bath, insulate, slowly expand. **Stage 3:** isothermal; insert apparatus in low $T = T_c$ bath, slowly compress. **Stage 4:** adiabatic; remove apparatus from bath, insulate, slowly compress.

behavior also reflects the fact that during a reversible adiabatic expansion, T decreases along with P (see Figure 13.1 and also Figure 13.2).

Power laws are defined in the Log Blog post on p. 83; γ is defined on p. 70.

13.2 Carnot Cycle & Absolute Zero Temperature

The *Carnot cycle* is *the* quintessential thermodynamic cycle, covered in every thermodynamics textbook. It provides a useful and quantitative understanding of the operation of heat engines (Section 12.1), whereas the *reverse* Carnot cycle does the same for heat *pumps* (e.g., air conditioners). It is also an excellent example for analyzing how entropy behaves under reversible change—in which context, it relates to the Third Law of Thermodynamics.

From the Texts: well, *almost* all of them, anyway...

particularly with regard to heat engine *efficiencies* (p. 110)...

In a Carnot cycle, the "loop" is a precise set of four consecutive but reversible stages, as indicated in Figure 13.2. Two of these stages are expansions, and two are compressions. Also, two are isothermal, and two are adiabatic. Being a cycle, for all state functions X, ΔX is zero around the entire cycle—but not necessarily zero over each individual stage. Finally, $W = -Q \neq 0$ around the entire cycle—with $(W < 0, Q > 0)$ corresponding to a clockwise traversal (heat engine), and $(W > 0, Q < 0)$ to the reverse or counterclockwise traversal (heat pump).

In practice, care must be taken to ensure that every stage is as reversible as possible. For example, heat flow must be avoided when the apparatus is inserted into the hot bath at state A (or cold bath at state C), by ensuring that T is equal to T_h (or T_c), at the end of the preceding adiabatic stage 4 (or 2).

The area under the expansion stage curves (1 and 2) in Fig. 13.2 is greater than that under the compression stage curves (3 and 4).

CARNOT CYCLE: Behavior of Thermodynamic Quantities

For each stage of the Carnot cycle, ΔX, Q, and W behave as follows:

Stage	Temperature	Heat	Entropy	Work
1	$T=T_h$	$Q=Q_h>0$	$\Delta S=Q_h/T_h>0$	$W<0$
2	$\Delta T=-(T_h-T_c)<0$	$Q=0$	$\Delta S=0$	$W<0$
3	$T=T_c$	$Q=Q_c<0$	$\Delta S=Q_c/T_c<0$	$W>0$
4	$\Delta T=(T_h-T_c)>0$	$Q=0$	$\Delta S=0$	$W>0$
Total	$\Delta T=0$	$Q=(Q_h+Q_c)$ >0	$\Delta S=(Q_h/T_h+Q_c/T_c)$ $=0$	$W=-(Q_h+Q_c)$ <0

Note: The above are general results that do *not* presume an ideal gas.

Note that Q_h is the absorbed heat, and $-Q_c = |Q_c|$ the released or "wasted" heat, as per Section 12.2. Their difference is the work performed by the system, $-W = Q_h - |Q_c|$. That $W < 0$ follows from the fact that more heat is absorbed than released ($Q_h > |Q_c|$). This in turn follows from $\Delta S = Q_h/T_h + Q_c/T_c = 0$, together with the fact that $T_h > T_c$.

the hot bath is hotter than the cold bath...

▹▹▹**Helpful Hint:** Pay close attention to the signs of quantities indicated in the previous discussion and the box above it. Note that "verbal" quantities such as *released heat* or *work performed by the system* are usually taken to be positive.

The equation for the entire cycle ΔS given above can be rewritten as

Note that $Q_c \to 0$ as $T_c \to 0$—a consequence that directly relates to the Third Law (Sec. 13.3).

$$\frac{|Q_c|}{Q_h} = \frac{T_c}{T_h}, \qquad \text{[Carnot cycle]} \quad (13.1)$$

which says that the wasted heat can be minimized by maximizing the temperature ratio between hot and cold heat baths. Conversely, the efficiency of the heat engine—or ratio of work performed to absorbed heat—is defined as $\eta = |W|/Q_h = (1 - T_c/T_h)$. Both this expression and Equation (13.1) are completely general—i.e., applicable to both ideal gases and other materials.

they can even be applied to other cycles...

Remarkably, η depends *only* on the temperature ratio, and not at all on the properties of the heat engine itself! The other side of this coin, though, is that for fixed T_c and T_h, η is always less than one, and cannot be increased. Only in the limit that $T_c \to 0$ does $\eta \to 1$. Note also that $T_c < 0$ would result in $\eta > 1$—an obviously physically incorrect prediction that reconfirms the absolute zero limit on temperature.

another depressing result of the Second Law...

From the Texts: In a formal sense, negative temperatures *do* exist! See **Atkins**, p. 591; **Kittel**, Appendix E; **Reif**, p. 105.

Real heat engines are even less efficient than predicted above. In part, this is due to some inevitable irreversibility, which always reduces $|W|$ (see

Sections 14.3 and 14.4). In part, though, also, this is by design; power stations, for example, are usually optimized for *power* (rate of work), rather than efficiency.

From the Texts: Baierlein, Sec. 3.3; **Callen**, Sec. 4.9; **Cengel**, Chaps. 6 and 10.

13.3 Third Law & Absolute Entropy

The final Law of Thermodynamics is the Third Law, which—like the Second Law—also deals with entropy. Our introduction to the Third Law will be empirical—like that of the Second Law in Section 12.1.

There are not a great many laws in thermodynamics.

Consider a material that can transform between two different condensed phases. The entropy change associated with this (reversible) phase transition is denoted '$\Delta_{trs}S$'. It is an experimental fact that in *all such cases*, $\Delta_{trs}S \to 0$ as $T \to 0$. Thus, both phases approach the same S value. If we take the $S(T=0)$ value to be zero, we obtain the

From the Texts: Two of the best treatments are: P. A. **Rock**, *Chemical Thermodynamics* (University Science Books, 2003), Chap. 6; **Reif**, Sec. 4 · 6.

Third Law: *The absolute entropy of the system approaches zero as the absolute temperature approaches zero:*

$$S \to 0 \quad \text{as} \quad T \to 0 \qquad \text{[always]} \quad (13.2)$$

Are we allowed, though, to simply "set" $S=0$? We are, *if* we use the thermodynamic definition of Equation (10.1)—according to which, S is defined only up to an additive constant. In this context, the Third Law thus converts *relative* to *absolute* entropies—essentially serving to define the additive constant for the latter.

See Secs. 10.1, 11.1, and 11.4. Absolute entropies are also called "Third-Law entropies."

Using the *information* Definition 10.3, however, absolute entropies are obtained directly (in principle at least)—and so the Third Law must be *proved*, not just assumed. The Boltzmann distribution of Equation (6.1) implies that as $T \to 0$, the system spends all of its time in just a *single* molecular state. Thus, $\Omega \to 1$ as $T \to 0$, and so according to Equation (10.2), $S \to 0$.

This is the lowest energy, or *ground* state. Even if there is some ground state *degeneracy* ($\Omega > 1$), k effectively reduces S to zero.

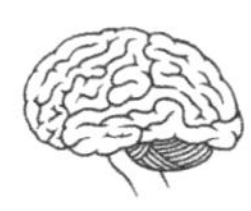

▷▷▷ **To Ponder...** The ΔS form of the Third Law is called the *Nernst theorem*; the S form is due to Planck. It could be argued that the former is the actual physical law, since S cannot be measured directly.

▷▷▷ ***Don't* Try It!!** Don't try to apply the Third Law to the Sackur-Tetrode expression [Equation (11.11)]—which predicts *negative infinite* S as $T \to 0$. This fails because the ideal gas law is invalid in the $T \to 0$ limit.

Part V

Free Energy

"Waste not free energy; treasure it and make the best use of it."

—Wilhelm Ostwald

"If to any homogeneous mass, we suppose an infinitesimal quantity of any substance to be added, the mass remaining homogeneous and its entropy and volume remaining unchanged, the increase of energy of the mass divided by the quantity of the substance added is the potential for that substance in the mass considered."

—J. Willard Gibbs

"Gibbs' prose style conveys his meaning…using no more than twice as many words as Poincaré or Einstein would have used to say the same thing."

—E. T. Jaynes

A Conceptual Guide to Thermodynamics, First Edition. Bill Poirier.

Companion website: http://www.conceptualthermo.com

Chapter 14

Free Energy & Exergy

Anderson: Sec. 4.3, Sec. 4.5, Sec. 4.8, Sec. 4.12, Sec. 4.13, Sec. 4.17; **Atkins:** Sec. 3.5, Sec. 3.7, Sec. 3.8; **Atkins-life:** Sec. 2.6, Sec. 3.2; **Baierlein:** p. 151, Sec. 7.2, Chap 10; **Callen:** Sec. 4-5, Sec. 5-3, Sec. 6-1, Sec. 6-2, Chap. 7; **Cengel:** Chap. 8, Sec. 12–2; **Chang:** p. 41, Sec. 4.6; **Elliott:** Sec. 4.12, Sec. 6.1, pp. 233–235; **Engel:** Sec. 6.1, Sec. 6.2; **Faure:** Sec. 11.7, Sec. 11.8; **Kittel:** pp. 68–72, pp 262–265; **Levine:** Sec. 4.3, Sec. 4.4; **McQuarrie:** Sec. 21–1, Chap. 22; **Moran:** Chap. 7, Sec. 11.3; **Prausnitz:** Sec. 2.1, Sec. 2.2, Sec. 3.1; **Reif:** Sec. 5 · 5, Sec. 5 · 6; **Sandler:** p. 110, pp. 192–194; **Schroeder:** Sec. 5.1; **Silbey:** p. 45, Sec. 4.1, Sec. 4.2; **Smith:** Sec. 6.1; **Tinoco:** pp. 72–80

▷▷▷ **Try It!!** Look up "free energy" on the internet. No doubt you will find mostly conspiracy-theory websites, claiming that "free" (meaning *ridiculously inexpensive*) sources of energy are plentifully available—if only Big Oil (or whomever…) would just let us get our hands on it!

That is *not* the kind of free energy we will be talking about here—but then again, it is not entirely unrelated, either. There is, in fact, a great deal of energy all around us. Most of this is not *free*, however—in the sense of being *available* to perform work.

which is why *useful* energy costs money…

If the Second Law teaches anything, it is that it is impossible to convert all of a system's internal energy into useful work. It is thus natural to define new thermodynamic quantities known as *free energies*—which represent that portion of U that *is* available to do work.

From the Texts: Indeed, free energy is sometimes referred to as *"availability."*

A Conceptual Guide to Thermodynamics, First Edition. Bill Poirier.

Companion website: http://www.conceptualthermo.com

14.1 What Would Happen If Entropy Were a Variable?

It all begins with the differential form of the First Law, Equation (8.1). Now that we also know about entropy, we can rewrite this [using Equation (10.1)] as

$$dU = TdS - PdV. \qquad \text{[reversible, infinitesimal]} \quad (14.1)$$

Equation (14.1) is remarkably simple and elegant. The right-hand side involves four quantities—each of which appears exactly once, three of which are known thermodynamic variables, and one of which is entropy.

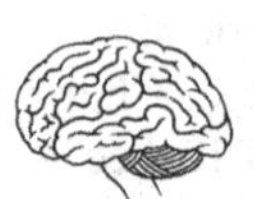

▷▷▷ **To Ponder...** When scientists see something simple and elegant like this, they call it "suggestive"—as if nature is revealing a sign or a hint...

In particular, Equation (14.1) seems to "scream" that S should be treated as a thermodynamic *variable*, not just a quantity. Indeed, treating $U = U(S, V)$ as a function of the two independent variables, S and V, Equation (14.1) becomes the *total differential* for U.

The idea that S is a variable may seem "weird," but ***Science Doesn't Care***...

Henceforth, the set of thermodynamic variables is taken to be (T, S, P, V), rather than (T, P, V). Only two of these four variables are still independent, however, as can be seen clearly in the $U(S, V)$ example above. The new picture nevertheless suggests a new choice of "natural" independent variables for U—i.e., (S, V), rather than the (T, V) set that has been used up till now (see Section 9.2). How do the natural variables change for the other thermodynamic quantities? Can we choose *any* two variables that we want, from our new set of four, to serve as independent variables? Looking more closely at Equation (14.1), it seems that the four variables come in two *pairs*: (T, S) and (P, V). The two variables in each pair are called *conjugate variables*.

at least for *closed* systems (i.e., until Chap. 15)...

From the Texts: Some authors prefer the term "proper independent variables."

CONJUGATE VARIABLES: Introduction

- Thermodynamic variables come in *conjugate pairs*.
- Each conjugate pair has one extensive and one intensive variable.
- Each conjugate pair corresponds to one kind of thermodynamic contact.
- The product of two conjugate variables has dimensions of energy.
- One variable from each conjugate pair is used in the set of independent variables.

The last bullet in the box above implies that there are a total of four possible choices for the set of independent variables—namely: (S, V); (S, P);

(T, V); (T, P). Uniquely associated with each of these four choices is a natural energy-like quantity—a so-called *thermodynamic potential*, obtained from U in similar fashion to the way that H was derived in Section 9.2. Specifically, the "trick" is to add (or subtract) conjugate pair products to U, in order to transform the natural variables appropriately.

Try It!! Mathematically, this is known as a *Legendre transform.*

THERMODYNAMIC POTENTIALS: Energy-like Quantities

There are four thermodynamic potentials—each an energy-like state function quantity, associated with one of the four choices of independent variables listed above:

Name	Symbol	Variables	Definition	Total Differential
Internal Energy	U	(S, V)		$dU = TdS - PdV$
Enthalpy	H	(S, P)	$U + PV$	$dH = TdS + VdP$
Helmholtz Free Energy	A	(T, V)	$U - TS$	$dA = -SdT - PdV$
Gibbs Free Energy	G	(T, P)	$U + PV - TS$	$dG = -SdT + VdP$

14.2 Helmholtz and Gibbs Free Energies

Like enthalpy, the two new *free energy* quantities, A and G, serve as practical tools. They enable the Second Law to be applied directly to the *system*, rather than the total system—thus making it possible to forget about the surroundings entirely, and once again proving that **It's OK to be Lazy**. This confers a great practical advantage in predicting which thermodynamic changes are spontaneous—but, as we will see shortly, only under certain conditions. As promised, the free energies also describe the extent to which the internal energy can be transformed into useful work.

Throughout this section, it may prove useful to consult the box above.

like the Clausius inequality of Equation (12.2)...

Note that the total differential for *every* thermodynamic potential in the box above has the same elegant form as Equation (14.1)—all four variables appear exactly once, and in conjugate pairs. Whereas V is associated with U, and P with H, as was the case in Section 9.2, both U and H are now associated with S, rather than T. Transforming U and H into new quantities that depend naturally on T rather than S, one obtains A and G, respectively—which is another way to motivate the introduction of the latter.

You may have noticed that P and V always seem to go together; the same is also true of T and S.

Again, one more chance to **Try It!!**

Note also that A and G are obtained by *subtracting* the conjugate variable product TS from U and H (respectively), rather than adding it. This is a critical point; since all four variables are positive, free energy values are always *less* than the corresponding energies. To some extent, this reflects the Second Law fact that only some of the energy can be converted into useful work.

though as a caveat, recall from the discussion on p. 4 that free energy can be negative...

Since the independent variables (T, P) are generally preferred to (T, V), G is more important than A, in practice. This is especially true for thermodynamic processes that occur at *constant* (T, P), such as phase transitions

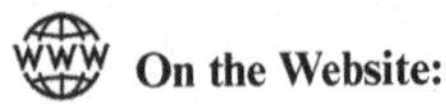
On the Website:
http://www.conceptualthermo.com

and chemical reactions (see Chapter 17 and the website). For this reason, we will often single out G for special treatment—even though similar analyses might hold for A, or even U and H.

For example, by comparing the total differential dG from the box above with Equation (9.16), we obtain the following elegant partial derivative relations:

$$\left(\frac{\partial G}{\partial T}\right)\bigg|_P = -S \quad ; \quad \left(\frac{\partial G}{\partial P}\right)\bigg|_T = V \tag{14.2}$$

Note the *minus sign* in front of the S in Equation (14.2) above—reflecting the idea that higher T leads to a greater *entropic* energy contribution that is "useless" for doing work.

Equation 14.2 also leads to an important partial derivative relation on the thermodynamic variables themselves—known as a *Maxwell relation.* Applying the cross-derivative trick of Section 9.4 [i.e., Equation (9.18)] to Equation (14.2),

$$-\left(\frac{\partial S}{\partial P}\right)\bigg|_T = \left(\frac{\partial V}{\partial T}\right)\bigg|_P . \qquad \text{[Maxwell relation]} \tag{14.3}$$

There are four such Maxwell relations in all, one obtained from each thermodynamic potential.

▷▷▷ **Try It!!** Try to derive all four Maxwell relations on your own, starting from each of the four thermodynamic potentials. Along the way, you will also derive six more partial derivative relations of the form of Equation (14.2). See also the Helpful Hint below.

Collectively, the set of four Maxwell relations describe the constraints imposed on the variables, (T, S, P, V)—by the fact that only two of these are independent. In this respect, they serve the same kind of role as the equation of state does for (T, P, V).

▷▷▷ **Helpful Hint:** Students are often asked to prove Maxwell relations. The first step is to confirm that the equation presented is a partial derivative relation involving *only* the four variables. The second step is to identify the two independent variables from the partial derivatives—e.g., T and P in Equation (14.3). The third step is to write down the total differential for the corresponding energy-like quantity, and to apply cross derivatives.

14.3 Second Law & Maximum Work

For a constant pressure change, $dQ = dH$ by Equation (9.9), and so Equation (12.2) becomes

$$dH - TdS < 0. \qquad \text{[irreversible, const } P\text{, infinitesimal]} \quad (14.4)$$

Assuming, also, that temperature is held constant, consider the following derivation:

DERIVATION: Gibbs Criterion for Spontaneous Change

$$dH - TdS - SdT < 0 \qquad [dT = 0]$$
$$d(H - TS) = dG < 0 \qquad [\text{definition of } G]$$
$$\therefore\ \Delta G < 0 \qquad [\text{irreversible, const } (T, P)] \quad (14.5)$$

Thus, the sign of ΔG of the *system* determines whether a process is spontaneous or not—provided that it takes place at constant (T, P). Likewise, if $\Delta G = 0$, then the process is reversible. Similar relations hold for ΔA, for processes at constant (T, V).

▷▷▷ **Helpful Hint:** Make sure to get the *direction* correct, when working with the various inequalities derived from the Second Law. In particular, $\Delta G|_{T,P} < 0$—but $\Delta S_{\text{tot}} > 0$—for a spontaneous irreversible change.

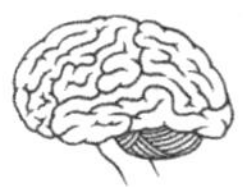

▷▷▷ **To Ponder...** For a pure substance in a single phase (Section 3.1), the two variables (T, P) [or (T, V)] *define* the thermodynamic state—which therefore *cannot change* if both variables are held fixed. It is important to realize that Equation (14.5) [and also the ΔA analog] applies *only* to more complicated situations such as phase transitions and chemical reactions.

On the other hand, ΔA *does* say something important about gas expansions—provided that these occur at constant T. Specifically, ΔA is equal to the reversible, or *maximum work* that can be extracted from a true $(\Delta V > 0)$ isothermal gas expansion:

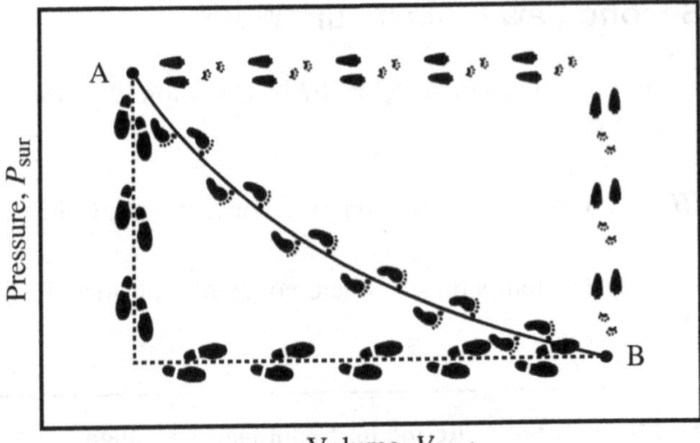

Figure 14.1 The path less traveled. Three travellers follow three different isothermal paths from A to B: A. Lincoln (boots) takes the lower, irreversible path; M. Ghandi (bare feet) takes the middle, reversible path; E. Bunny (paw prints) takes the upper, irreversible path. The last path is *not* spontaneous and therefore forbidden by the Second Law—*the Easter Bunny doesn't exist!*

DERIVATION: Maximum Work

$$dA = -SdT - PdV \quad \text{[total differential]}$$
$$dA = -SdT + dW = dW \quad \text{[reversible, } dT=0\text{]}$$
$$\therefore \ \Delta A = W \quad \text{[reversible, const } T\text{]} \quad (14.6)$$

See also Fig. 13.1. The dark gray vertical plane in that figure corresponds to the isothermal plane of Fig. 14.1.

Figure 14.1 expands on the discussion at the end of Section 8.4, by offering additional insight as to why the reversible isothermal path provides maximum work. Any isothermal path lying above the reversible path would indeed generate more work; however, all such paths are nonspontaneous, and thus forbidden by the Second Law.

Note that *directionality matters*. For the *reverse* process—i.e., the B-to-A *compression* process—the *upper* irreversible path is spontaneous, and the *lower* irreversible path is forbidden. The reversible work now becomes the *minimum work* that must be applied to bring about the compression.

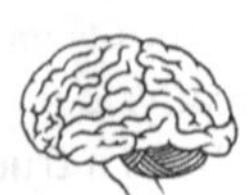

▷▷▷ **To Ponder...** There is also a maximum *non-expansion* work principle, based on ΔG. This version requires constant (T, P)—and therefore only applies to more complicated situations, as per the To Ponder on p. 119. Examples include chemical reactions, and *exergy* (see below).

14.4 Exergy

From Section 13.2 we know that work can be generated whenever there is access to two heat baths at different temperatures—simply by sandwiching a heat engine in between them. This strategy succeeds because the heat is not transferred directly from the hot bath to the cold bath, but rather through an *intermediary* that is always at the same temperature as the bath it is in contact with. Even so, only some fraction of the absorbed heat is converted into work—a fraction that diminishes with the temperature difference, moreover.

at least in principle...

This is important to know, because eventually, even large heat baths will thermally equilibrate with each other—once sufficient heat has been transferred via repeated heat engine cycles. In reality, the "isothermal" stages of the Carnot cycle are not truly so; because of heat transfer, each cycle reduces T_h and increases T_c, however slightly—until eventually, $T_h = T_c$, and no more work may be extracted. Useful work may also be obtained from two bodies that manifest a *pressure* difference—again, most efficiently through the agency of a reversible intermediary. Here, too, however, useful work ceases as soon as mechanical equilibrium of the two bodies is achieved.

In reality, the "cycle" of Fig. 13.2 is a slowly collapsing spiral.

In all then, how much useful work can be extracted before thermal and mechanical equilibrium are both reached? The theoretical maximum is known as the *exergy*, denoted 'E'. Exergy is *nearly* equivalent to Gibbs free energy, as can be understood by considering the following important points:

This is *not* the same question that was posed and answered in Sec. 13.2—i.e., "how much work can be obtained from a *single* heat engine cycle?"

1. Exergy requires *three* bodies, rather than two (system, surroundings, and intermediary.)
2. The intermediary enables a transfer of energy that is *isothermal, isobaric*, and *reversible*.
3. Under such conditions, ΔG is equal to the maximum *useful* or "non-expansion" work.

The implied equivalence will be clarified below (see also To Ponder on p. 120).

Exergy can be formulated as follows. First, define the "system" to be the hot bath, rather than the intermediary (with $T = T_h$, as per Section 13.2). Second, take the (presumed) much larger cold bath to be the surroundings. Typically, the cold bath is the atmosphere, in which case $T_c = T°$ and $P_c = P°$ take on *standard values*. Consequently, $T°$ and $P°$ remain effectively constant throughout the process, and useful work may be generated only until the system reaches $T = T°$ and $P = P°$—the so-called *dead state*.

both baths are large compared to the intermediary...

Standard (ambient) values are denoted using the 'o' superscript; specific $T°$ and $P°$ values are given in Sec. 4.4.

Exergy is the useful work released by the *total system*, as the system progresses from its initial state to the dead state. Because of reversibility, $\Delta S_{tot} = Q_{tot} = 0$ (also, $\Delta V_{tot} = 0$). Because of the intermediary, however, the total system is *not* isolated, and so the *useful work*, $W_{tot} = \Delta U_{tot} = \text{E}$, is nonzero. Note that W_{tot} excludes the "useless" expansion work done on the surroundings by the system (or vice-versa), and is therefore equivalent to the "non-expansion" work.*

*though this may be a misnomer; the form of W_{tot} is unspecified, and thus could well be expansion work, performed against a *fourth* body...

DERIVATION: Exergy

$$dU_{sur} = T^\circ dS_{sur} - P^\circ dV_{sur} \qquad \text{[total differential]}$$
$$\Delta U_{sur} = T^\circ \Delta S_{sur} - P^\circ \Delta V_{sur} \qquad \text{[const } (P^\circ, T^\circ)]$$
$$\Delta U_{sur} = -T^\circ \Delta S + P^\circ \Delta V \qquad [\Delta S_{tot} = \Delta V_{tot} = 0]$$
$$\therefore \; E = \Delta U_{tot} = \Delta U - T^\circ \Delta S + P^\circ \Delta V \qquad [\Delta U_{tot} = \Delta U + \Delta U_{sur}] \qquad (14.7)$$

The similarity between E and G is now obvious, as is the key difference—because of its dependence on T° and P°, exergy is a combined property that depends on *both* the system *and* the surroundings.

In practice, engineers also include other forms of energy in Equation (14.7), such as kinetic and potential energy of the system relative to the surroundings, chemical energy, electrical energy, etc. It may be that the exergy concept can be generalized still further.

From the Texts: For a *highly* engaging discussion of the role of entropy and exergy in our daily lives, consult **Cengel**, pp. 347–349, pp. 465–468.

Chapter 15

Chemical Potential, Fugacity, & Open Systems

Anderson: Sec. 4.7, Sec. 4.14, Sec. 7.5, Sec. 8.1, Sec. 8.3; **Atkins:** p. 127, pp. 129–130, p. 137, Sec. 5.1; **Atkins-life:** Sec. 3.7, Sec. 3.8(a); **Baierlein:** Chap. 7, Chap. 10; **Callen:** Sec. 2-1, Sec. 2-8, Sec. 18-3, p. 414; **Cengel:** Sec. 1–3, pp. 709–712; **Chang:** pp. 108–109, p. 131, pp. 200–201; **Elliott:** Sec. 4.14; **Engel:** Sec. 6.3, Sec. 6.4, Sec. 7.5; **Faure:** Sec. 11.8, Sec. 11.9; **Kittel:** pp. 117–134, p. 139; **Levine:** Sec. 4.1, Sec. 4.6, Sec. 6.1; **McQuarrie:** Sec. 22–8, Sec. 23–3; **Moran:** Sec. 1.2.2, Sec. 11.9.2, Sec. 11.9.3, Sec. 11.9.4; **Prausnitz:** Sec. 1.2, Sec. 2.2, Sec. 2.3, Sec. 2.4, Sec. 2.6. Sec. 2.7, Appendix A; **Reif:** Sec. 8 · 7, Sec. 8 · 10; **Sandler:** Sec. 7.4, Sec. 8.2; **Schroeder:** Sec. 1.1, Sec. 3.5, Sec. 3.6, pp. 156–158, pp. 161–166; **Silbey:** Chap. 4; **Smith:** Sec. 11.1, Sec. 11.2, Sec. 11.6; **Tinoco:** pp. 87–93, pp. 101–104, p. 144

15.1 What Would Happen If n Were a Variable?

In the early part of this book, the amount of substance, n (or N) was treated as a thermodynamic *variable*. Since then, we have for simplicity considered only closed systems, for which n is constant. This is fine for describing simple processes such as gas expansions. However, most thermodynamic applications of real-world interest—phase transitions, mixtures, chemical reactions, osmosis, thermoelectric power generation, etc. (see Chapter 17 and the website)—are not so simple. Such processes involve *open* systems—or those for which n can otherwise change.

i.e., up to the start of Sec. 4.1...

On the Website:
`http://www.conceptualthermo.com`

Engineers refer to open systems as *control volumes*.

How does this affect our thermodynamic picture as developed thus far? Regarding n as a thermodynamic variable, the most basic question is whether or not this variable is *independent*. Since it is clearly possible to change the thermodynamic state of a system by adding new particles to it—*without* changing either of the two variables (T, P)—n must indeed be independent.

You do this every time you fill your water bottle—for which "filled" and "empty" states are most definitely macroscopically distinct!

A Conceptual Guide to Thermodynamics, First Edition. Bill Poirier.

Companion website: http://www.conceptualthermo.com

For an open system consisting of a pure substance in a single phase, there are thus three independent variables in all. In this context, n can only change as a result of *diffusion*—i.e., the random migration of particles, in or out of the system. The system and surroundings are said to be in *diffusive contact*, separated by a diffusive or permeable wall.

The *most* diffusive wall is no wall at all—which is certainly permitted.

Though important in its own right, our present study of diffusion is mainly motivated by the desire to set the stage for more complicated processes—specifically, phase transitions, mixtures, and chemical reactions. In such cases, the system is comprised of more than one *type* of particle—thus requiring *multiple* independent n variables. Moreover, the particle types can interconvert—allowing for n values to change even if the system is closed. Though quite involved, a solid understanding of these more advanced topics rests firmly on the foundations laid in this chapter.

Recall that according to Section 14.1, thermodynamic variables come in *conjugate pairs*. This means that the variable n ought to have a conjugate partner—some new variable we have not yet considered. What should be the properties of this hypothetical new variable? Assuming that **It's OK to be Lazy**, the boxes in Section 14.1 practically define the new quantity for us.

Note that μ has *nothing* to do with the *potential energy*, E_P, of Chap. 5.

Let the *chemical potential*, μ, denote the thermodynamic variable conjugate to n. Since n is extensive, μ must be *intensive*—with dimensions of energy per mole. The three conjugate pairs are thus (T, S), (P, V) and (μ, n). One independent variable can be chosen from each pair—although in practice, n is almost always used from the last pair.

The thermodynamic variables and their key relationships are summarized in the box below; each row corresponds to a different conjugate pair:

CONJUGATE VARIABLES: Summary

Intensive variable	Extensive variable	Conjugate product	Contact/ equilibrium	Energy transferred
T	S	TS	thermal	Q
P	V	PV	mechanical	W
μ	n	μn	diffusive	$\mu \Delta n$

Furthermore, with n now treated as a variable, we expect new, generalized total differentials for the thermodynamic potentials, of the following form:

THERMODYNAMIC POTENTIALS: Revised for Variable n

Thermodynamic potential	Total differential
$U(S, V, n)$	$dU = TdS - PdV + \mu\, dn$
$H(S, P, n)$	$dH = TdS + VdP + \mu\, dn$
$A(T, V, n)$	$dA = -SdT - PdV + \mu\, dn$
$G(T, P, n)$	$dG = -SdT + VdP + \mu\, dn$

With $dn = 0$, these total differentials reduce to those of the second box in Sec. 14.1, as is appropriate.

15.2 Chemical Potential

From the discussion in Section 15.1—and especially the two boxes—the meaning of the chemical potential is now clear. Specifically, from the lower right corners of these two boxes, we obtain the following:

Definition 15.1 *The* chemical potential, *denoted 'μ', measures the change in Gibbs free energy with respect to a change in n, at constant (T, P):*

$$\mu = \left(\frac{\partial G}{\partial n}\right)\bigg|_{T,P} \qquad \text{[always]} \quad (15.1)$$

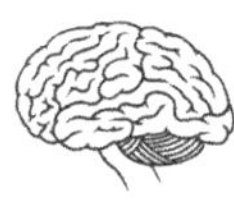

▷▷▷ **To Ponder...** Note that μ could just as easily be defined using any of the other three thermodynamic potentials, U, H, or A. However, G is most useful, in part because both of the constant variables are intensive.

The extensivity of G implies $G(T, P, n) = nG_m(T, P)$. It thus follows from Definition 15.1 that $\mu = G_m$. So why not take $\mu = G_m$ to be the definition—or simply dispense with μ altogether, and work with G_m instead? This strategy would only work for a pure substance in a single phase; in more general situations, μ and G_m are *not* equivalent [although Equation (15.1) always holds].

To understand the role that μ plays in diffusion, consider a system that is suddenly brought into diffusive contact with its surroundings. In perfect analogy with thermal and mechanical contact, $\mu \neq \mu_{sur}$ implies a lack of *diffusive equilibrium*—leading to a spontaneous net flow of particles from high μ to low μ. This flow continues until $\mu = \mu_{sur}$, and equilibrium is restored.

perhaps via sudden removal of the dividing wall...

Just like for the other two types of equilibria, the above situation can be interpreted in entropic/information terms. In accord with Section 12.4, diffusive contact removes a macroscopic restriction on the system—i.e., the system is no longer closed. The system then takes advantage of its new freedom by exploring the expanded range of molecular states now available to it—i.e., particles migrate between the system and surroundings. This results in a macroscopic change of state—i.e., a change in the variable n. Macroscopic change continues until the total system entropy is maximized, at which point diffusive equilibrium is restored. It is the same basic thermodynamic drama, reenacted yet again, but with a different cast of players...

Remember that diffusion is a model for more complex μ-driven processes such as phase transitions and chemical reactions—for which equal μ values *also* signal equilibrium.

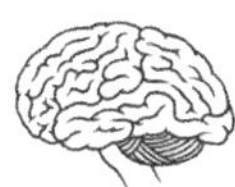

▷▷▷ **To Ponder...** Even after the system has reached diffusive equilibrium, particles continue to drift randomly in and out. Thus, $N = N(t)$ is not strictly constant over time, but fluctuates—much like $E(t)$ in Figure 5.1. In analogy with U, then, the true thermodynamic quantity for open systems is actually $\langle N(t) \rangle$, rather than N itself. However, many authors (including this one!) are "sloppy," by using 'N' to denote both N and $\langle N(t) \rangle$. In practice, the two quantities are very nearly identical for sufficiently large systems—e.g., those of macroscopic, $N \approx N_A$ proportions.

15.3 Ideal Gas & Fugacity

Consider the following derivation of ΔG, for an ideal gas under reversible expansion at constant (T, n):

DERIVATION: ΔG for Ideal Gas at Constant (T, n)

$$dG = VdP \qquad \text{[total differential, const } (T,n)]$$

$$dG = nRT\left(\frac{dP}{P}\right) \qquad \text{[Eq. 4.5 (ideal gas law)]}$$

$$\therefore \Delta G = \int_{P_i}^{P_f} nRT\left(\frac{dP}{P}\right) = nRT\ln\left(\frac{P_f}{P_i}\right) \qquad \text{[ideal gas, const } (T,n)] \tag{15.2}$$

Dividing Equation (15.2) by n results in a similar expression for the ideal gas isothermal $\Delta\mu$. This form is useful for relating the μ of a standard reference state (see Section 14.4), to some unknown μ at a different pressure. Specifically, taking $P_f = P$ as the pressure of interest, and taking the initial state to be the standard reference state ($P_i = P^\circ$), we obtain the following:

Here, $\mu = \mu(T, P)$, and $\mu^0 = \mu(T, P^\circ)$—and very often, $T = T^\circ$ (Sec. 14.4).

$$\mu = \mu^\circ + RT\ln\left(\frac{P}{P^\circ}\right) \qquad \text{[ideal gas]} \tag{15.3}$$

Note that Equation (15.3) provides a simple relation between μ and P, for ideal gases. This relation is extremely important in practice, because it enables a determination of μ directly from P—which is typically much easier to measure in a laboratory experiment.

Of course, Equation (15.3) breaks down for *non-ideal* systems. Nevertheless, even in the non-ideal case, it would be convenient to work with some pressure-*like* quantity that *does* directly relate to μ. This motivates the following definition—for the *fugacity*, f, or "tendency to flee":

like Harrison Ford in *The Fugitive*...

$$\mu = \mu^\circ + RT\ln\left(\frac{f}{P^\circ}\right) \qquad \text{[fugacity]} \tag{15.4}$$

Thus, f is what replaces P, to make Equation (15.3) true in the general, non-ideal case.

By design, fugacity has dimensions of pressure, and also $f = P$ for the ideal gas. For non-ideal systems, $f < P$ (or $f > P$) if attractive (or repulsive) intermolecular interactions dominate. Thus, fugacity plays a similar role to the compressibility factor, Z (p. 30), and the internal pressure, π_T (Section 9.3), in that it informs us the extent to which—and direction in which—a substance is non-ideal.

CORRESPONDENCE: Quantities & Intermolecular Interactions

Compressibility factor	Internal pressure	Fugacity	Intermolecular interactions
$Z = 1$	$\pi_T = 0$	$f = P$	none (ideal gas)
$Z < 1$	$\pi_T > 0$	$f < P$	attractions dominate
$Z > 1$	$\pi_T < 0$	$f > P$	repulsions dominate

The main role of fugacity, however, is to serve as a pressure-like substitute for the chemical potential. In fact, Equation (15.4) implies that $f = f_{\text{sur}}$ is a valid condition for diffusive equilibrium (provided $T = T_{\text{sur}}$ and $P = P_{\text{sur}}$). Most engineers (including chemical engineers!) prefer f to μ, particularly when applying these quantities to equilibrium phase transitions and mixtures (phase equilibria).

One minor weakness of the Equation (15.4) definition of fugacity is that it depends on the choice of reference state, and is therefore only a "relative" quantity. Physicists prefer to work with a dimensionless, "absolute" fugacity quantity, defined as $\exp(\mu/RT)$.

From the Texts: For an explanation of why, see **Sandler**, p. 407.

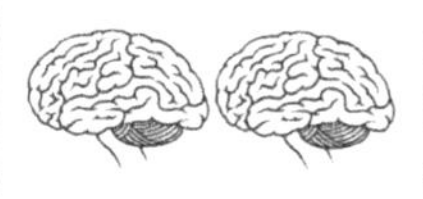

⊳⊳⊳ **To Ponder...*at a deeper level.*** The absolute fugacity has direct relevance for statistical mechanics—particularly for *quantum* statistical phenomena (e.g., Bose-Einstein condensation) whose onset occurs when $\exp(\mu/RT) \approx 1$ (see, e.g., **Callen**, Section 18-3).

The absolute fugacity is also called the *absolute activity*, by some authors. It is closely related to the *activity* in solution chemistry, and to the *fugacity coefficient* in the study of phase equilibria—both of which are also dimensionless quantities. However, one important difference is that the latter are *relative* quantities, despite being dimensionless.

From the Texts: See, e.g., **Kittel**, p. 139.

Part VI

Applications

"Man's greatness consists in his ability to do and the proper application of his powers to things needed to be done."

—Frederick Douglass

"Only one thing registers on the subconscious mind: repetitive application—practice. What you practice is what you manifest."

—Fay Weldon

"Adultery is the application of democracy to love."

—H. L. Mencken

"There are very few personal problems that cannot be solved through a suitable application of high explosives."

—Scott Adams, creator of *Dilbert*.

A Conceptual Guide to Thermodynamics, First Edition. Bill Poirier.

Companion website: http://www.conceptualthermo.com

Chapter 16

Crazy Gay-Lussac's Gas Expansion Emporium

Anderson: Sec. 3.4.1, Sec. 3.5, Sec. 4.4.1, Sec. 4.11, Sec. 4.12, Sec. 4.13; **Atkins:** Sec. 2.3, Sec. 2.4, Sec. 2.5, Sec. 2.6, Sec. 3.2, Sec. 3.3, Sec. 3.5, Sec. 3.9; **Atkins-life:** Sec. 1.3, Sec. 1.4, Sec. 1.6, Sec. 2.2, Sec. 2.6; **Baierlein:** Sec. 1.3, Sec. 1.5, pp. 37–39, Sec. 2.6, Sec. 3.7, Sec. 7.2, Sec. 10.1, Sec. 10.5, Sec. 10.6; **Callen:** Sec. 3-4, Sec. 4-1, Sec. 4-2, Sec. 4-5, pp. 120–121, Sec. 6-3, Sec. 7-4; **Cengel:** Sec. 2–6, Sec. 4–1, p. 195, Sec. 7–1, Sec. 7–4, Sec. 7–5, Sec. 7–9, Sec. 12–4; **Chang:** Sec. 3.1, Sec. 3.4, Sec. 4.2, Sec. 4.3, Sec. 4.4, Sec. 4.6, Sec. 4.8; **Elliott:** Sec. 2.4, Sec. 2.8, Sec. 2.10, pp. 74–87, Sec. 4.1, Sec. 4.3, Sec. 4.4, Sec. 4.5, Sec. 9.2, Sec. 9.4; **Engel:** Sec. 2.8, Sec. 2.9, Sec. 2.10, Sec. 2.11, Sec. 5.4, Sec. 5.6, Sec. 5.7, Sec. 6.1, Sec. 6.3; **Faure:** Sec. 11.2, Sec. 11.3, Sec. 11.6, Sec. 11.7, Sec. 11.8; **Kittel:** pp. 171–176, pp. 236–246; **Levine:** Sec. 2.4, Sec. 2.5, Sec. 2.8, Sec. 2.9, Sec. 2.12, Sec. 3.4, Sec. 3.5, Sec. 4.2, Sec. 4.3, Sec. 4.5; **McQuarrie:** Sec. 19–1, Sec. 19–2, Sec. 19–4, Sec. 19–5, Sec. 19–7, Sec. 20–3, Sec. 20–4, Sec. 20–6, Sec. 22–1, Sec. 22–2; **Moran:** Sec. 2.2, p. 56, Sec. 3.12, Sec. 3.15, Sec. 5.11, Sec. 6.5, Sec. 6.11, Sec. 11.4; **Prausnitz:** n/a; **Reif:** Sec. 2 · 10, Sec. 3 · 6, Sec. 3 · 9, Sec. 3 · 11, Sec. 4 · 1, Sec. 4 · 5, Sec. 5 · 1, Sec. 5 · 3, Sec. 5 · 4, Sec. 5 · 5, Sec. 5 · 9; **Sandler:** Sec. 3.1, Sec. 3.2, Sec. 3.4, Sec. 4.1, Sec. 4.2, Sec. 4.4, Sec. 4.5, Sec. 6.4; **Schroeder:** pp. 25–26, Sec. 1.2, Sec. 1.5; **Silbey:** Sec. 2.7, Sec. 2.10, Sec. 3.2, Sec. 3.3, Sec. 4.2, Sec. 4.3, Sec. 4.4; **Smith:** Sec. 2.8, Sec. 2.9, Sec. 2.10, Sec. 5.5, Sec. 5.7, Sec. 5.8, Sec. 6.1; **Tinoco:** p. 20, pp. 30–36, pp. 48–49, pp. 60–62, pp. 70–77

16.1 Sales Pitch

You want gas expansions? Crazy Gay-Lussac's got 'em—in all models, shapes, and sizes, and so *cheap he's practically* giving *them away. So come on down to Crazy Gay-Lussac's Gas Expansion Emporium today!*

Even if the above hype does not exactly get you excited about them, gas expansions are indeed the oldest and most universal application of

it seems that "hot air" is the *product* here, *as well* as the marketing gimmick...

A Conceptual Guide to Thermodynamics, First Edition. Bill Poirier.

Companion website: http://www.conceptualthermo.com

thermodynamics. Moreover, they are still very relevant today: still used routinely in engineering practice; still taught in virtually every thermodynamics course regardless of discipline; still embodying core thermodynamics principles. Accordingly, gas expansions are the one from a myriad of applications that appears in this book (but see Chapter 17...).

As alluded to above and in Section 8.5, gas expansions *do indeed* come in a great many flavors—so many that when solving problems, it can be difficult to know where to begin. Part of our job in this chapter is thus to sort through all of this variety for you, thereby serving as a (hopefully) useful reference. In particular, Section 16.3 provides a "comprehensive compendium" of increasingly specific results, presented in outline form.

On the other hand, our larger mission (as always) is to convey a sense of how to approach problems *systematically*. Section 16.2 thus provides the tools needed to tackle *any* gas expansion, without resorting to memorization of all the special cases.

Recall the discussion on p. xi of the Preface, as well as the second Don't Try It on p. 6, and the Helpful Hint on p. 7.

▹▹▹ **Helpful Hint:** While reading this chapter, you may find it helpful to refer to certain previous material, such as: Fig. 7.1 (p. 50); Sec. 7.2 (especially Helpful Hints); Fig. 8.1 (p. 58); Don't Try Its on p. 59; box on p. 60; Fig. 8.2 (p. 61); Sec. 8.5; Sec. 9.2; Log Blog posts on p. 63 and p. 83; Fig. 13.1 and surrounding discussion (p. 108).

16.2 How to Solve Gas Expansion Problems

The "zeroth" step in solving a gas expansion problem is to identify the *type* of gas expansion that it is—with respect to the following four *attributes*:

TYPES OF GAS EXPANSIONS: Four Attributes

- reversible vs. irreversible
- free vs. non-free
- isothermal vs. isobaric vs. adiabatic
- ideal vs. non-ideal

All of these attributes or conditions have been defined previously—e.g., in Section 8.5.

With one choice for each of the four attributes in the box above, 24 distinct gas expansion types emerge in all—though in practice, not all of these are realized. Also, certain types (e.g., reversible, non-free, isothermal, ideal) are seen much more frequently—the *everyday specials*!

The remaining steps are outlined below; these need not be followed exactly, but serve as a guideline:

SOLVING GAS EXPANSION PROBLEMS: Six Steps

1. Choose independent variables (first = V; second = whatever is constant).
2. Determine initial and final states (as needed; using equation of state).
3. Calculate ΔX values [for desired state functions; using Equation (7.1)].
4. Calculate W by integrating along the path [Equation (8.4), or one of the conditional forms of Section 8.5].
5. Calculate Q from ΔU and W [if needed; using Equation (8.5)].
6. Calculate ΔS_{sur} by integrating $dS_{\text{sur}} = -dQ/T$ (box on p. 105).

In Step 1, the second variable should obviously be T for an isothermal expansion, or P or P_{sur} for an isobaric expansion; in both cases, the constant nature of the variable should be exploited to make the problem easier to solve. Steps 2 and 3 above use the fact that for both the initial and final states, the system is in equilibrium (Section 8.4). This also implies that $P_{\text{sur}}(V_i) = P_i$ and $P_{\text{sur}}(V_f) = P_f$. This is true for *all* gas expansions (including irreversible ones). Step 4 is the only integration that is usually required; from W, the quantities Q and ΔS_{sur} can be obtained, in Steps 5 and 6, respectively.

The adiabatic case is discussed below—as are various other common special cases that we have not yet discussed much.

Adiabatic expansion: This case is tricky, because all three variables can change. The strategy is thus to exploit $Q=0$—leading, e.g., to $\Delta S_{\text{sur}}=0$, and to a simple expression for work that does not require integration ($W=\Delta U$). These steps are easier than for the isothermal and isobaric cases, but Step 2 is harder.

Reversible adiabatic expansion of ideal gas: Here, Step 2 can be achieved by following an adiabat, $P(V) \propto V^{-\gamma}$ (Section 13.1, Figure 13.1, and Figure 13.2), where the adiabat coefficient γ is the heat capacity ratio (p. 70). The proof requires Equation (11.9), whose generalization for rotating ideal gases is as follows:

$$S(T, V) = nR\ln(CVT^{d/2}) = nR\ln\left[CV\left(\frac{PV}{nR}\right)^{d/2}\right] \quad \text{[ideal gas (rotating)]} \tag{16.1}$$

All ideal gas results in this chapter refer to *general* (rotating) ideal gases.

Note that d is the total number of coordinates per molecule, including the rotational coordinates (see discussion and box on p. 40). Since S is constant

(Section 11.5), so must be $V(PV)^{d/2}$, and also $PV^{(d+2)/d}$. Finally, from the box on p. 69 we can identify

$$\gamma = \left[\frac{d+2}{d}\right], \qquad \text{[ideal gas (rotating)]} \quad (16.2)$$

by also recognizing, from Equations (4.5), (5.8), and (9.6), that

$$H(T,P) = \left[\frac{d+2}{2}\right] nRT. \qquad \text{[ideal gas (rotating)]} \quad (16.3)$$

Isobaric expansion: Generally, this condition can be regarded as $P_{\text{sur}} = \text{const} = P_f$—the gas is expanded against a constant surroundings pressure equal to the final system pressure, resulting in $W = -P_f \Delta V$. For **reversible** isobaric expansion, $P_i = P = P_{\text{sur}}$, and T changes in accord with the equation of state. More interesting is **irreversible** isobaric expansion—arising (for true expansion) when $P_i > P_{\text{sur}}$, after the system and surroundings are brought into mechanical contact. The irreversible case can be—and often is—also *isothermal*. If so, then *all system state function ΔX values are identical to those of the corresponding reversible isothermal expansion*, since the initial and final states for the two paths are identical. This can save time on problems.

This is the irreversible path in Fig. 8.2.

Free expansion: This occurs when the partition between two chambers—one containing gas and the other empty—is suddenly ruptured or removed. Since $P_{\text{sur}} = W = 0$ (p. 58), the expansion ends *not* when $P_f = P_{\text{sur}}$, but rather, when the gas reaches the expanded limits of the container. All free expansions are true expansions ($\Delta V > 0$), and also irreversible. They can be **isothermal** or **adiabatic**—with these two special cases being identical to each other for the ideal gas ($\Delta T = 0 = \Delta U = Q + W$).

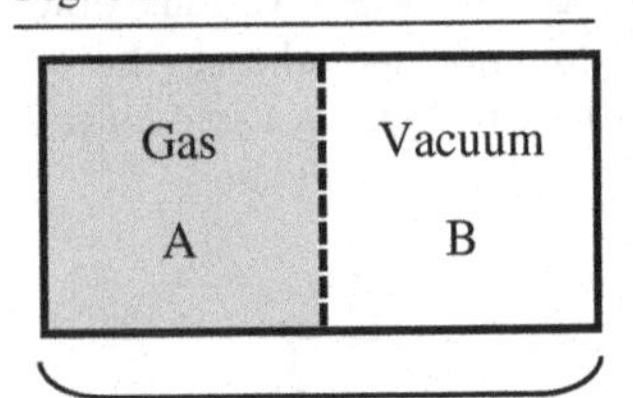

Removing partition (vertical dashed line) allows gas to flow from A to B.

Ideal gas: There are many important results that can be derived explicitly for the ideal gas. Many of these have already been provided in this book, for the ideal gas of point particles. As this chapter deals with *rotating* ideal gases however, we need more general expressions. In addition to those presented above, the following equations—derived from Equations (5.8) and (16.3), respectively—will also prove useful, particularly for the compendium (Section 16.3):

From the Texts: Additional results may be found in your primary textbook. An excellent source of reference material is H. D. B. **Jenkins**, *Chemical Thermodynamics at a Glance* (Blackwell, 2008). (See also Try It below.)

$$\Delta U = \frac{d}{2} nR\Delta T \qquad \text{[ideal gas (rotating)]} \quad (16.4)$$

$$\Delta H = \left[\frac{d+2}{2}\right] nR\Delta T \qquad \text{[ideal gas (rotating)]} \quad (16.5)$$

16.3 Comprehensive Compendium

The following *comprehensive compendium* of gas expansions is presented in outline form, with more specific conditions corresponding to greater indentation. To use the compendium, first identify the particular type of gas expansion, as per the box on p. 132. Then scan downward until you reach an attribute (e.g., "**Reversible**"): if it applies, keep scanning down; otherwise, jump to the next attribute that appears at the same indentation level. All entries encountered in this fashion will apply to your particular gas expansion.

⊳⊳⊳**Try It !!** Try to derive the compendium results on your own! Each entry includes a brief description: in some cases, this is all you need to derive the result; in other cases, you may need to do a bit more work…

GAS BLOG: When Solving Gas Expansion Problems…

⊳⊳⊳**Helpful Hint:** Always check to see if your final answers make sense. For typical macroscopic dimensions ($n \approx 1$ mol), the orders of magnitude for various quantities should be as follows: for energies, (W, ΔU, etc.), 10^3 J; for pressure, 10^5 Pa; for entropy (ΔS), 10^1 J/K (see Don't Try It and Helpful Hint on p. 20). The sign of W should be as follows: negative for a true expansion ($\Delta V > 0$); positive for a compression ($\Delta V < 0$).

⊳⊳⊳**Helpful Hint:** Use the state function property to avoid doing extra work in ΔX calculations. If the problem asks you to consider more than one path from the same initial state A to the same final state B, then ΔX need only be computed for one of the paths. Choose the simplest.

Gas expansion:

- $\Delta X = (X_f - X_i) \quad (\Delta X_{\text{tot}} = \Delta X + \Delta X_{\text{sur}})$ [X is any (extensive) state function]
- $\Delta V > 0 \quad (\Delta V < 0)$ [true expansion (compression)]
- $\Delta U = W + Q$ [First Law, Eq. (7.3)]
- $W = -\int_{V_i}^{V_f} P_{\text{sur}}(V)\, dV$ [definition, Eq. (8.4)]
- $W < 0 \quad (W > 0)$ [true expansion (compression)]
- $dS_{\text{sur}} = -dQ/T$ [surroundings in equilibrium with themselves; box, p. 105]

Reversible:

- $P = P_{\text{sur}}$ [mechanical equilibrium throughout]
- $W = -\int_{V_i}^{V_f} P(V)\, dV$ ["reversible work," Eq. (8.6)]
- $Q = \int_{S_i}^{S_f} T(S)\, dS \quad ; \quad dS = dQ/T$ [Eqs. (10.1) and (11.16)]
- $\Delta S_{\text{tot}} = 0 \quad ; \quad \Delta S = -\Delta S_{\text{sur}}$ [p. 107]

Isothermal:

- $\Delta T = 0 \quad ; \quad T_i = T_f = T = \text{const}$ [definition of isothermal]
- $W = \Delta A = \Delta U - T\Delta S$ [max work, Eq. (14.6); definition, p. 117]
- $Q = T\Delta S$ [box, p. 110]
- $\Delta G = \Delta H - T\Delta S$ [definition, p. 117]

Ideal gas:

- $\Delta U = \Delta H = 0$ [Eq. (16.4); Eq. (16.5)]
- $W = \Delta A = -nRT \ln(V_f/V_i)$ [Eq. (8.7)]
- $\Delta G = nRT \ln(P_f/P_i) = -nRT \ln(V_f/V_i)$ [box, p. 117]
- $Q = -W = nRT \ln(V_f/V_i)$ [definition, Eq. (8.5)]
- $\Delta S = -\Delta S_{\text{sur}} = nR \ln(V_f/V_i)$ [Eq. (11.5)]

Isobaric:

- $\Delta P = 0 \quad ; \quad P_i = P_f = P = \text{const}$ [definition of isobaric]
- $\Delta T > 0 \quad (\Delta T < 0)$ [true expansion (compression)]
- $W = -P\Delta V$ [P can be "pulled out" of the Eq. (8.6) integral]
- $Q = \Delta H = \Delta U + P\Delta V$ [Eq. (9.10); definition, Eq. (9.6)]

Ideal gas:

- $(T_f/T_i) = (V_f/V_i)$ ["Charles's Law" form of Eq. (4.5)]
- $\Delta U = (d/2)nR\,\Delta T$ [Eq. (16.4)]
- $W = -nR\,\Delta T$ [Eq. (4.5)]
- $Q = \Delta H = [(d+2)/2]nR\,\Delta T$ [Eq. (16.5)]
- $\Delta S = -\Delta S_{\text{sur}} = [(d+2)/2]nR \ln(T_f/T_i)$ [Eqs. (11.13) and (16.1)]
- $\Delta(A/T) = -(d/2)nR \ln(T_f/T_i)$ [box, p. 117; Eq. (16.4)]
- $\Delta(G/T) = -[(d+2)/2]nR \ln(T_f/T_i)$ [box, p. 117; Eq. (16.5)]

Adiabatic:

- $Q = 0$ [definition of adiabatic]
- $\Delta T < 0 \quad (\Delta T > 0)$ [true expansion (compression)]
- $\Delta U = W$ [First Law, Eq. (7.3)]
- $\Delta S = -\Delta S_{\text{sur}} = 0$ [discussion, p. 95]

Ideal gas:

- $\gamma = (C_P/C_V) = [(d+2)/d]$ [Eq. (16.2)]
- $P_i V_i^{\gamma} = P_f V_f^{\gamma}$ [discussion, p. 108]
- $T_i V_i^{\gamma-1} = T_f V_f^{\gamma-1}$ [Eq. (4.5)]

Gas expansion (cont.):

Reversible (cont.):

Adiabatic (cont.):

Ideal gas (cont.):

$\Delta U = W = (d/2)nR\,\Delta T$ [Eq. (16.4)]

$\Delta H = [(d+2)/2]nR\,\Delta T$ [Eq. (16.5)]

$\Delta A = (d/2)nR\,\Delta T - S\Delta T$ [box, p. 117; Eq. (16.4)]

$\Delta G = [(d+2)/2]nR\,\Delta T - S\Delta T$ [box, p. 117; Eq. (16.5)]

Irreversible:

$\Delta S_{\text{tot}} > 0 \quad ; \quad \Delta S > -\Delta S_{\text{sur}}$ [Second Law, Eq. (12.1)]

$dS > dQ/T$ [Clausius inequality, Eq. (12.2)]

Isothermal:

$\Delta T = 0 \quad ; \quad T_i = T_f = T = \text{const}$ [definition of isothermal]

$\Delta X = \Delta X$ for reversible isothermal [X is any state function]

$|W| < |\Delta A| \quad (|W| > |\Delta A|)$ [max work, true expansion (compression)]

$\Delta S_{\text{sur}} = -(Q/T)$ [box, p. 105; box, p. 110]

Isobaric (surroundings only):

$P_f = P_{\text{sur}} = \text{const}$ [expand against P_{sur} until mechanical equilibrium]

$P_i > P_{\text{sur}} \quad (P_i < P_{\text{sur}})$ [true expansion (compression)]

$W = -P_f \Delta V$ [P_{sur} can be "pulled out" of the Eq. (8.4) integral]

Isothermal and Isobaric:

$|P_f \Delta V| < |\Delta A| \quad (|P_f \Delta V| > |\Delta A|)$ [true expansion (compression)]

Ideal gas:

$Q = -W = P_f \Delta V$ [definition, Eq. (8.5); Eq. (16.4)]

$\Delta S_{\text{sur}} = -(P_f \Delta V)/T$ [$\Delta S_{\text{sur}} = -(Q/T)$]

Free:

$\Delta V > 0$ [true expansion only (no compression)]

$P_{\text{sur}} = 0$ [expand against $P_{\text{sur}} = 0$ until $V = V_f > V_i$]

$W = 0$ [$W \neq 0$ requires *both* $\Delta V \neq 0$ *and* $P_{\text{sur}} > 0$]

Isothermal:

$\Delta T = 0 \quad ; \quad T_i = T_f = T = \text{const}$ [definition of isothermal]

$\Delta X = \Delta X$ for reversible isothermal [X is any state function]

Ideal gas:

$Q = \Delta U = 0$ [definition, Eq. (8.5)]

$\Delta S_{\text{sur}} = 0$ [$\Delta S_{\text{sur}} = -(Q/T)$]

Adiabatic:

$Q = 0$ [definition of adiabatic]

$\Delta U = 0$ [First Law, Eq. (7.3)]

$\Delta S_{\text{sur}} = 0$ [$\Delta S_{\text{sur}} = -(Q/T)$]

Ideal gas:

$\Delta T = \Delta U = 0$ [Eq. (16.4)]

$\therefore$ identical to free isothermal expansion of ideal gas

Chapter 17

Electronic Emporium: *Free Online Shopping!*

Thank you for shopping at the Gas Expansion Emporium! We hope that you found everything you were looking for. If not, please be sure to visit our online *shopping venue, the* Electronic Emporium. *There, you will find an incredible variety of only the finest-quality goods, tailored to meet your specific shopping needs.*

The goals of this book as described in the Preface and Chapter 1—i.e., to provide a clear exposition of the *core concepts* of thermodynamics, pertinent across all relevant disciplines—have hopefully now been reached.

On the other hand, where the many *applications* of thermodynamics are concerned, one size does *not* always fit all. To be sure, everyone should know about gas expansions. However, not *absolutely* everyone needs to understand the intricacies of how quantum spin properties, as described by the two-dimensional square-lattice Ising model, can lead to ferromagnetic phase transitions and negative temperatures...

Accordingly, all discipline-specific application material is provided on the website—in the form of *additional book chapters*, available for download. At the time of printing, these include—as a bare minimum—chapters addressing three of the most important thermodynamics applications: *phase transitions*; *mixtures*; *chemical reactions*. As already anticipated, these common (albeit not universal) applications build from the foundation laid in Chapter 15—and of course, all of the previous book material, as well.

On the Website: http://www.conceptualthermo.com

Depending on student interest, author time, the availability of coauthors, etc., it is expected that a number of additional chapters will also eventually be written, addressing such potential topics as: amorphous solids; Bose-Einstein condensation; heat exchangers and power cycles; mass-dependent isotope fractionation; osmosis; protein folding; statistical mechanics; steam engines.

and perhaps even the aforementioned Ising model...

All of the online application chapters refer extensively to this book—as well as to the relevant reference textbooks from the set listed in the Textbook Guide. The additional chapters also refer to *each other*, but to a more limited extent. In cases where one chapter requires others as "prerequisites,"

A Conceptual Guide to Thermodynamics, First Edition. Bill Poirier.

Companion website: http://www.conceptualthermo.com

all such chapter dependencies are listed explicitly at the start of the chapter in question.

The policy at the time of printing allows you to download any of the additional chapters *for free*. This requires you to have: (a) purchased a print or ebook version of this book; (b) registered on the website. *But hurry! This offer is valid only while supplies last!*

Remember that registration also opens up other resources—such as textbook-specific materials, and references to other textbooks (or other editions) not listed in the Textbook Guide. It also offers a means for you to provide feedback—though you can also email comments directly to `feedback@conceptualthermo.com`. Feel free, for instance, to recommend new application chapters that you might like to see written.

or that you would like to *help write*, if you are a prospective author...

Part VII

Appendices

"Chaos often breeds life, when order breeds habit."

—Henry Adams

"The 'universal tendency towards dissipation' places entropic heat loss in the same semantic category as deplorable personal habits."

—Katherine Hayles

"Love is like the human appendix. You take it for granted while it's there, but when it's suddenly gone you're forced to endure horrible pain that can only be alleviated through drugs."

—Reverend Jen, *Live Nude Elf*

A Conceptual Guide to Thermodynamics, First Edition. Bill Poirier.

Companion website: http://www.conceptualthermo.com

Appendix A

Beards Gone Wild! Facial Hair & the Founding Fathers of Thermodynamics

This is not a book on the history of thermodynamics; nevertheless, it can sometimes be instructive to match names to faces. Below, we present portraits for several of the primary players in this field during its early days (i.e., the 19th and early 20th centuries), together with a brief description of their most important contributions.

Enjoy the panoply of magnificent minds and terrific topiary!

Ludwig Boltzmann: Father of statistical mechanics. Statistical definition of entropy (Boltzmann formula); Boltzmann distribution; entropy interpretation of the Second Law. Boltzmann constant (physical constant). Bipolar; died by own hand.

Sadi Carnot: Father of thermodynamics. Uncle of French president, Sadi Carnot. Sadly, no beard (though the nephew's is nice...) Heat engines (Carnot cycle, heat engine efficiency); Second Law. Died at 36, suffering from mania and delirium.

A Conceptual Guide to Thermodynamics, First Edition. Bill Poirier.

Companion website: http://www.conceptualthermo.com

Rudolf Clausius: Father of entropy (thermodynamic definition). First Law; entropy interpretation of the Second Law (Clausius inequality); phase transitions (Clausius-Clapeyron equation). Clausius unit of entropy.

Pierre Duhem: Important contributions to phase transitions and mixtures (Duhem theorem, Gibbs-Duhem equation). Respectable mane—though nothing compared to that of present-day beard champion with same name (not a joke!).

J. Willard Gibbs: Key contributions across *many* areas. Statistical mechanics (ensembles, Gibbs paradox); free energy (Gibbs free energy, Gibbs-Helmholtz equation); phase transitions and mixtures (phase rule, Gibbs-Duhem equation).

James Prescott Joule: Brewer. Father of heat and the First Law. Absolute temperature; internal pressure (Joule experiment); isenthalpic expansion (Joule-Thomson effect). SI unit of energy.

Lord Kelvin (William Thomson): Member of the UK House of Lords. First, Second, *and* Third Laws. Absolute temperature; isenthalpic expansion (Joule-Thomson effect); thermoelectricity (Thomson effect). SI unit of temperature.

James Clerk Maxwell: *Huge* name in physics. Statistical mechanics (Maxwell and Maxwell-Boltzmann distributions); entropy and Second Law (Maxwell's demon); free energy (Maxwell relations); phase transitions (Maxwell construction).

Walther Nernst: Father of the Third Law. Chemical reactions (chemical affinity); electrochemistry (Nernst equation). Inventor of the Nernst electric lamp, and the electric piano. Ran afoul of the Nazis in Germany, prior to his death in 1941.

William Rankine: First, Second, *and* Third Laws. Heat engines (Rankine cycle, similar to Carnot cycle); phase transitions (enthalpy of vaporization). Inventor of the term, "potential energy." Rankine unit of temperature.

Appendix B

Thermodynamics, Abolitionism, & Sha Na Na

In this appendix, we follow a rather convoluted path whose initial state is Josiah Willard Gibbs (1839–1903). J. W. Gibbs was an American scientist and Yale University professor whose monumental legacy is entirely out of proportion to his unassuming manner. In addition to his great achievements in thermodynamics and statistical mechanics (Appendix A), Gibbs made other profound contributions—e.g., in vector calculus and optics. He has been called "the greatest mind in American history" by no less a great mind than Albert Einstein himself [1]. One might well wonder whether Einstein ever had occasion to learn about other early American luminaries such as Benjamin Franklin, Thomas Jefferson, and James Madison…but certainly, his admiration for Gibbs cannot be in doubt.

Gibbs was rather less well appreciated by his own peers, however, at least at first. In part, this was due to the "curse of interdisciplinarity"—with chemists finding his work too mathematical, and vice-versa. In part, also, this may have been due to geography. Gibbs published his (now famous) discovery of the chemical potential in the *Transactions of the Connecticut Academy of Arts and Sciences* [2]—not exactly on the "must read" list for his European contemporaries such as Ludwig Boltzmann. Finally, it must be said that Gibbs may also have been a victim of his own verbose but subdued style—alluded to earlier, in the quote by E. T. Jaynes on p. 113.

▷▷▷ **Try It !!** Go back and reread the J. W. Gibbs quote that directly precedes the E. T. Jaynes quote on p. 113. Can you deciper which quantity Gibbs is talking about?

(**Answer:** chemical potential)

Gibbs seems to have inherited his quiet, non-rock-star demeanor from his father—another Josiah Willard Gibbs, also a Yale faculty member. A linguist, theologian, and abolitionist, the senior Gibbs played a vital role in the *Amistad* incident—a fascinating precursor to the American Civil War, and focus of the eponymous 1997 movie directed by Steven Spielberg. In 1839, African slaves on a Spanish ship out of Cuba—the *Amistad*—successfully mutinied. The ship eventually arrived on American shores (Long Island, specifically) from which the Africans were sent to a jail in New Haven until such time as their status could be sorted out. Thus began two years of legal wrangling that would wend its way up to the US Supreme Court. At issue was not only the fate of the Amistad Africans, but also the future course of slavery in America.

A successful legal battle would require testimony by the Africans—none of whom spoke English. The linguist Gibbs ascertained that their native language was *Mende*, but could only speak a few words of that

A Conceptual Guide to Thermodynamics, First Edition. Bill Poirier.

Companion website: http://www.conceptualthermo.com

language himself. He conceived a brilliantly innovative plan to find a Mende interpreter: wander around the New Haven and New York docks counting from one to ten in Mende, until recognized by a native speaker. In this manner, Gibbs found James Covey—a sailor from an anti-slavery British patrol ship who was more than willing to serve this (not so covertly) abolitionist cause.

A successful legal battle would also require a master attorney at the helm. Such was finally found in John Quincy Adams—the former US President, who at the time was the US Representative from Massachusetts' 12th district. Already wrestling against "gag rule" legislation that banned the mere mention of the word "slavery" in Congress—yet not fully trusted by abolitionists either, whose tactics he disagreed with—the aging Adams delivered a masterful performance before the US Supreme Court on behalf of the Amistad Africans [3]. Adams won the case and the Africans were set free—in what was arguably the most significant salvo against slavery yet seen in the US. Indeed, Adams' young congressional colleague from Illinois, a certain Abraham Lincoln (Figure 14.1), would borrow heavily from the Adams playbook, twenty years later.

John Quincy Adams was also an ardent believer in federal support for the advancement of science, who helped to establish the Smithsonian Institution and the US Naval Observatory. Decades later, his grandson Henry would become famous for his own writings on scientific and technological advances [4,5]—especially on the profound social impact of the rapid pace of progress. Not a scientist himself, Henry Adams' fascination with thermodynamics nevertheless led to attempts to incorporate First and Second Law principles into other disciplines such as history. Alas, as Adams himself lamented, he was unable to achieve this goal with anything approximating thermodynamical rigor, owing to the fact that he was not really a "math guy." In retrospect, Henry should probably not have been so hard on himself: history is inherently harder than thermodynamics; people are not molecules.

Be that as it may, there is no question that thermodynamic ideas—particularly the Second Law and its end-of-the-world implication of final "heat death" (p. 100)—kept a strong grip on the popular imagination of the late 19th and early 20th centuries, entering into the culture in myriad ways. Both scientists and nonscientists employed colorful language to characterize the inevitable loss of useful energy—"fatigue," "dissipation," "degradation," etc. (see From the Texts, p. 122). Such terminology seems to have resonated with other Victorian concerns—suggesting an even broader, allegorical application of thermodynamic concepts to social science and moral philosophy (p. 141). A detailed and engaging treatment is presented in Bruce Clarke's book, *Energy Forms* [6]. From H. G. Wells to D. H. Lawrence to Matt Groening (creator of *The Simpsons*)—from Morlocks to Lady Chatterley to Akbar and Jeff—the broad and long lasting cultural ramifications of thermodynamic ideas are revealed, with insight and a touch of understated humor.

Oh yes, and Bruce "Bruno" Clarke was the original bass player for the gold-lamé-sporting, 1950s-inspired rock band, *Sha Na Na*.

References

1. "J. Willard Gibbs," *Physics History*, American Physical Society, http://www.aps.org/programs/outreach/history/historicsites/gibbs.cfm.
2. J. W. Gibbs, "On the Equilibrium of Heterogeneous Substances," *Transactions of the Connecticut Academy of Arts and Sciences*, **3**, 108–248, 343–524 (1874–1878).
3. Harlow Giles Unger, *John Quincy Adams* (Da Capo Press, 2012).
4. Henry Adams, *The Education of Henry Adams* (Houghton Mifflin, 1918).
5. Henry Adams,"A Letter to American Teachers of History," *The Degradation of the Democratic Dogma* (Macmillan, 1920).
6. Bruce Clarke, *Energy Forms: Allegory and Science in the Era of Classical Thermodynamics* (University of Michigan Press, 2001).

Appendix C

Thermodynamics & the Science of Steampunk

Although all in favor of clear and simple definitions, we are not going to even try to define *steampunk*. This is a topic that even the afficionadoes argue over—prompting, e.g., a six-and-a-half minute YouTube video just to answer the basic question of what it is [1]. Clearly still evolving faster than any mere definition can keep up with, there are nevertheless certain telltale thematic elements that have emerged from the steampunk movement, which are likely to persist into the future.

One of these elements, certainly, would be a fascination with the Victorian era (late 19th century)—both its technology and its society, but above all, its own *vision of the future*. This brings us to the second key element of the steampunk movement—imagining an alternate or parallel reality that might have been, but was not. To this end, the movement draws inspiration from contemporary authors such as Jules Verne and H. G. Wells. As we have alluded to in Appendix B, however, those authors themselves drew inspiration from thermodynamics.

Although the visionary authors of the 19th century certainly predicted great technological advances for the 20th century, they could not have known the *direction* that these would take [2]—nor the profound social changes that these would engender (Appendix B). The ultra convenient, fast-paced, mass-produced lifestyle of the 20th century is thus eschewed in steampunk—in favor of technology that is more idiosyncratic and personalized, and requires time, craftsmanship, and ingenuity to master. This would be a third common steampunk element. Within these broad parameters, steampunk has proven itself to be very expansive and highly creative—encompassing science, history, technology, and design, as well as theatre, costume, literature, and music.

As for the *science* of steampunk, this really is largely thermodynamics—with some electromagnetism thrown in for good measure. Steam engines, after all, are the *raison d'être* for thermodynamics (p. 1) [3]. They are also sufficiently cumbersome, temperamental, and inefficient as to be largely impractical on anything approximating the personal scale—which happens to dovetail nicely with the steampunk aesthetic. Though steam *cars* (p. 98) were in fact a reality—with the famous "Stanley Steamer" even enjoying a very brief market dominance over its internal combustion rivals [4]—steam-powered ray guns, goggles, and personal transport devices are another matter.

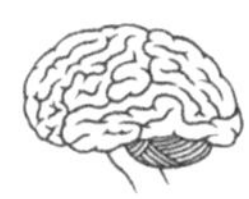

▷▷▷ **To Ponder...** It could be argued that Victorians *really were* partial to impracticality—certainly in terms of the formality of their social conventions, if not also their love of complex new technologies. *Bathing machines*—sea-side contraptions wheeled out into the water (sometimes under steam power!) to ensure proper modesty—are a

A Conceptual Guide to Thermodynamics, First Edition. Bill Poirier.

Companion website: http://www.conceptualthermo.com

classic example. Another case in point is Lewis Carroll. Most famous as the author of *Alice in Wonderland*, Carroll was also a highly accomplished practitioner of the new art and science of photography—until it developed to the point of being less-than-nearly-impossible to master, after which it seems he lost interest. Carroll, incidentally, thought bathing machines were ridiculous...

There is another, extremely important thermodynamics-related science that occasionally finds its way into steampunk: *information theory*. Though not formally established as its own discipline until the mid-20th century [5], information theory clearly has 19th century roots. In addition to Boltzmann's statistical definition of entropy (Definition 10.3, p. 84), the world's first *computer* had already been invented by Charles Babbage by the mid-19th century. Babbage's "difference engine" was a *mechanical computer*, designed to do surprisingly sophisticated numerical calculations. Later, while working with Ada Lovelace (the world's first computer programmer) to develop the even more sophisticated "analytical engine," the team also pioneered punch cards and "printers."

References

1. Antonia Sophie, *SteamPunk What Is It & a Request* (YouTube video, 2009) http://www.youtube.com/watch?v=smnCoV2wKnw.
2. One very interesting example is Jules Verne's little read *Paris in the Twentieth Century* (Del Rey, 1997), which has some *very* scarily prescient "near misses," as well as other predictions that are widely off the mark.
3. If there were sufficient interest, I could be persuaded to write an online application chapter on the operation of steam engines.
4. Nowadays, the name *Stanley Steemer* has been coopted by a chain of carpet cleaners.
5. C. Shannon, "A Mathematical Theory of Communication," *Bell System Technical Journal*, **27**, 379–423 (1948).

Steampunk Gallery

Goggles, top hat, ray gun, check.

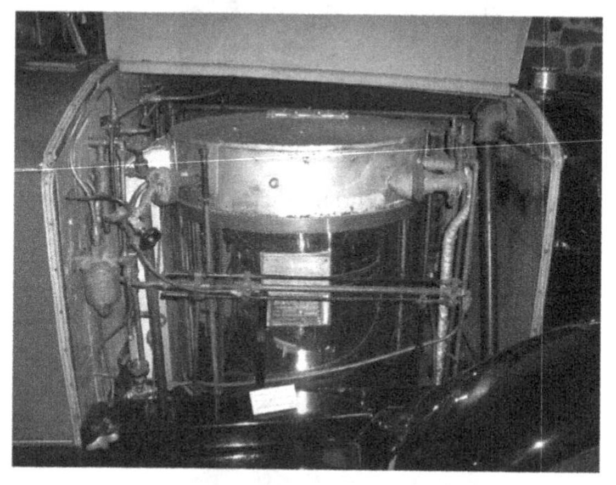

Stanley Steamer with boiler.

Gorgeous, painstaking craftsmanship is a hallmark of steampunk.

What if Charles Babbage had also invented the computer keyboard?

A Conceptual Guide to Thermodynamics, First Edition. Bill Poirier.

Companion website: http://www.conceptualthermo.com

Steampunks seem to go in big for disembodied brain imagery…

Charles Babbage's actual brain.

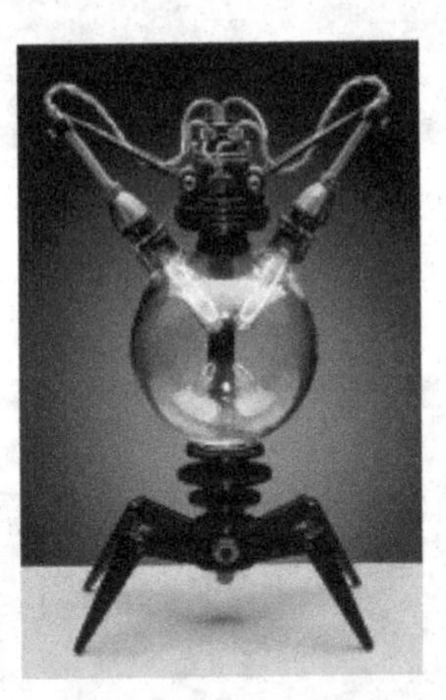

Does it move as well as light up?

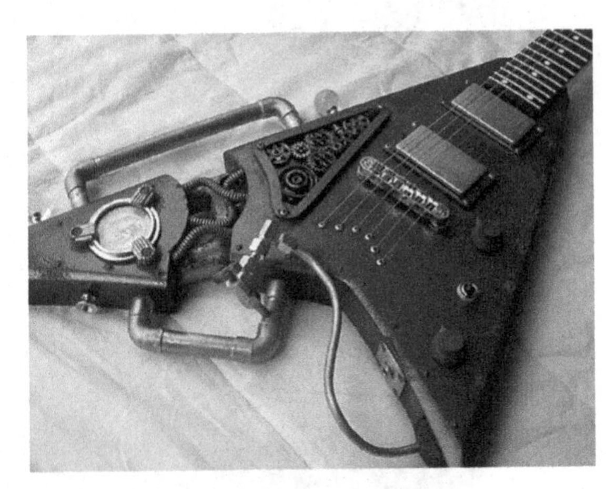

Nigel Tufnel's steampunk guitar?

Travel Try Its

▷▷▷ **Travel Try It!!** Cumbres & Toltec narrow gauge steam train, wending its way between Chama, New Mexico and Antonito, Colorado. Steam engines tend to "overpressurize" rather than overheat—thus needing to "blow off steam" occasionally, as pictured here (though the docent confessed that they also like to do it because it "looks cool.")

▷▷▷ **Travel Try It!!** Wyddfa steam train, chugging up from Llanberis station to the top of the incredibly moody and atmospheric Mount Snowdon (alleged home of Merlin the wizard) in northwest Wales. Note that it *pushes* rather than pulls—an innovation to deal with the extreme grade.

▷▷▷ **Travel Try It!!** Boltzmann's grave, across from a bunch of musicians, in the famous Zentralfriedhof cemetery in Vienna. The Boltzmann formula is prominently displayed. Note the following new entry for the Terminology and Notation Key (p. xviii):

Boltzmann Grave: ln→log; $\Omega \rightarrow W$

A Conceptual Guide to Thermodynamics, First Edition. Bill Poirier.

Companion website: http://www.conceptualthermo.com

Photo Credits

Beards Gone Wild:

Boltzmann: Wikimedia Commons / Public Domain
Carnot: Photogravure by Meisenbach Riffarth & Co. / Wikimedia Commons / Public Domain
Clausius: Heliogravure by Meisenbach, Riffarth & Co. / Wikimedia Commons / Public Domain
Duhem: Wikimedia Commons / Public Domain
Gibbs: Wikimedia Commons / Public Domain
Joule: C. H. Jeens / Wikimedia Commons / Public Domain
Kelvin: Sir Hubert von Herkomer / Wikimedia Commons / Public Domain
Maxwell: Wikimedia Commons / Public Domain
Nernst: Wikimedia Commons / Public Domain
Rankine: William Rankine / Wikimedia Commons / Public Domain

Steampunk Gallery:

watchman: Steampunk Watchman / steampunk-pics.com
Stanley Steamer: user:Liftarn / Wikimedia Commons / Public Domain
ceiling lamp: Photo by the author.
keyboard: The "von Slatt Original" Keyboard / Courtesy of "Datamancer" / datamancer.com
brain jar: Steampunk Brain in a Jar / Courtesy of Guy Garrison
Babbage brain: Alan Levine / Wikimedia Commons / CC-BY-2.0
machine light: Machine Light Type No. 01 / Courtesy of Frank Buchwald / frankbuchwald.de
guitar: Rusty Beauty / Courtesy of Jeff Ritzmann / thundereagleguitars.com

Travel Try Its: All photos by the author.

A Conceptual Guide to Thermodynamics, First Edition. Bill Poirier.

Companion website: http://www.conceptualthermo.com

Nigel Tufnel:	This is a top to a…you know, what we use on stage, but it's very…very special because if you can see…
Marty DiBergi:	Yeah…
Nigel Tufnel:	…the numbers all go to eleven. Look…right across the board.
Marty DiBergi:	Ahh…oh, I see…
Nigel Tufnel:	Eleven…eleven…eleven…
Marty DiBergi:	…and most of these amps go up to ten…
Nigel Tufnel:	Exactly.
Marty DiBergi:	Does that mean it's…louder? Is it any louder?
Nigel Tufnel:	Well, it's one louder, isn't it? It's not ten. You see, most…most blokes, you know, will be playing at ten. You're on ten here…all the way up…all the way up…
Marty DiBergi:	Yeah…
Nigel Tufnel:	…all the way up. You're on ten on your guitar… where can you go from there? Where?
Marty DiBergi:	I don't know…
Nigel Tufnel:	Nowhere. Exactly. What we do is, if we need that extra…push over the cliff…you know what we do?
Marty DiBergi:	Put it up to eleven.
Nigel Tufnel:	Eleven. Exactly. One louder.
Marty DiBergi:	Why don't you just make ten louder and make ten be the top…number…and make that a little louder?
Nigel Tufnel:	*[pause]* These go to eleven.

Index

Note: Page numbers that are underlined refer to definitions, laws, and theorems; those that are *italicized* refer to figures and marginal notes; those that are **boldfaced** refer to (gray) boxes.

A Conceptual Guide to Thermodynamics, First Edition. Bill Poirier.

Companion website: http://www.conceptualthermo.com

Printed in the United States
By Bookmasters